SPRINGER PROCEEDINGS IN PHYSICS 123

SPRINGER PROCEEDINGS IN PHYSICS

113 **Theoretical and Numerical Unsaturated Soil Mechanics**
Editor: T. Schanz

114 **Advances in Medical Engineering**
Editor: T.M. Buzug

115 **X-Ray Lasers 2006**
Proceedings of the 10th International Conference, August 20–25, 2006, Berlin, Germany
Editors: P.V. Nickles, K.A. Janulewicz

116 **Lasers in the Conservation of Artworks**
LACONA VI Proceedings, Vienna, Austria, Sept. 21–25, 2005
Editors: J. Nimmrichter, W. Kautek, M. Schreiner

117 **Advances in Turbulence XI**
Proceedings of the 11th EUROMECH European Turbulence Conference, June 25–28, 2007, Porto, Portugal
Editors: J.M.L.M. Palma and A. Silva Lopes

118 **The Standard Model and Beyond**
Proceedings of the 2nd International Summer School in High Energy Physics, Mǧla, 25–30 September 2006
Editors: M. Serin, T. Aliev, N.K. Pak

119 **Narrow Gap Semiconductors 2007**
Proceedings of the 13th International Conference, 8–12 July, 2007, Guildford, UK
Editors: B. Murdin, S. Clowes

120 **Microscopy of Semiconducting Materials 2007**
Proceedings of the 15th Conference, 2–5 April 2007, Cambridge, UK
Editors: A.G. Cullis, P.A. Midgley

121 **Time Domain Methods in Electrodynamics**
A Tribute to Wolfgang J. R. Hoefer
Editors: P. Russer, U. Siart

122 **Advances in Nanoscale Magnetism**
Proceedings of the International Conference on Nanoscale Magnetism ICNM-2007, June 25–29, Istanbul, Turkey
Editors: B. Aktas, F. Mikailov

123 **Computer Simulation Studies in Condensed-Matter Physics XIX**
Editors: D.P. Landau, S.P. Lewis, and H.-B. Schüttler

124 **EKC2008 Proceedings of the EU-Korea Conference on Science and Technology**
Editor: S.-D. Yoo

125 **Computer Simulation Studies in Condensed-Matter Physics XX**
Editors: D.P. Landau, S.P. Lewis, and H.-B. Schüttler

126 **Vibration Problems ICOVP 2007**
Editors: E. Inan, D. Sengupta, M.M. Banerjee, B. Mukhopadhyay, and H. Demiray

127 **Physics and Engineering of New Materials**
Editors: D.T. Cat, A. Pucci, and K.R. Wandelt

Volumes 90–112 are listed at the end of the book.

D.P. Landau S.P. Lewis H.-B. Schüttler
(Eds.)

Computer Simulation Studies in Condensed-Matter Physics XIX

Proceedings of the Nineteenth Workshop
Athens, GA, USA, February 20–24, 2006

With 59 Figures

Springer

Professor David P. Landau, Ph.D.
Professor Steven P. Lewis, Ph.D.
Professor Heinz-Bernd Schüttler, Ph.D.
Center for Simulational Physics
The University of Georgia
Athens, GA 30602-2451, USA

Springer Proceedings in Physics ISSN 0930-8989

ISBN 978-3-642-09922-9 e-ISBN 978-3-540-85625-2

DOI 10/2007.978-3-540-85625-2

Bibliographic information published by Die Deutsche Bibliothek
Die Deutsche Bibliothek lists this publication in the Deutsche Nationalbibliografie; detailed bibliographic data is available in the Internet at <http://dnb.ddb.de>.

Cover concept: eStudio Calamar Steinen
Cover design: SPI Publisher Services

Printed on acid-free paper

9 8 7 6 5 4 3 2 1

springer.com

Preface

Two decades ago, because of the tremendous increase in the power and utility of computer simulations, The University of Georgia formed the first institutional unit devoted to the use of simulations in research and teaching: The Center for Simulational Physics. As the international simulations community expanded further, we sensed a need for a meeting place for both experienced simulators and neophytes to discuss new techniques and recent results in an environment which promoted lively discussion. As a consequence, the Center for Simulational Physics established an annual workshop on Recent Developments in Computer Simulation Studies in Condensed Matter Physics. This year's workshop was the nineteenth in this series, and the continued interest shown by the scientific community demonstrates quite clearly the useful purpose that these meetings have served. The latest workshop was held at The University of Georgia, February 20–24, 2006, and these proceedings provide a "status report" on a number of important topics. This volume is published with the goal of timely dissemination of the material to a wider audience.

We wish to offer a special thanks to IBM for partial support of this year's workshop.

This volume contains both invited papers and contributed presentations on problems in both classical and quantum condensed matter physics. We hope that each reader will benefit from specialized results as well as profit from exposure to new algorithms, methods of analysis, and conceptual developments.

Athens, GA, USA
July 2006

D. P. Landau
S. P. Lewis
H.-B. Schüttler

Contents

Part V Biological and Soft Condensed Matter

1

Computer Simulation Studies in Condensed Matter Physics: An Introduction

D.P. Landau, S.P. Lewis, and and H.-B. Schüttler

Center for Simulational Physics, The University of Georgia,
Athens, GA 30602-2451, USA

Computer simulation studies in condensed matter physics now play a fundamental role in many areas of investigation. The "status report" which is contained in this volume is the result of presentations and discussion that occurred during the 19th Annual Workshop at the Center for Simulational Physics. The texts of both longer, invited presentations as well as a number of contributed papers are included. The reader will find that the scope of simulational/computational studies is broad and that substantial potential for cross-fertilization of methods between different sub-fields is evident.

The volume opens with four papers on materials properties. First, Fichthorn and Miron address the challenges associated with simulating rare events, where many time scales may be relevant. They use the hyperdynamics approach to accelerate dynamical simulation of such systems in a manner that correctly samples the equilibrium state of the fast processes while evolving the trajectories at a time scale relevant for the slow processes. This is achieved by on-the-fly consolidation of broad basins of high-frequency localized states into coarser states that incorporate the equilibrium properties of the smoothed-away minima. The authors demonstrate the technique on a realistic model system of the diffusion of Co clusters on the Cu (001) surface. Dennis and Liebig study the intensity dependence of the reflectance of three types of dielectric interference filters. They simulate the reflectance of both low- and high-intensity pulses using the finite-difference time-domain method and incorporating nonlinear optical behaviors in the field equations. The three systems are compared to assess the robustness of their reflective properties against nonlinear effects at extremely high intensities. Abraham extends his earlier work on crack instability in brittle fracture by applying a recently discovered scaling law to a model solid with a crack constrained to unidirectional travel. He finds that suppression of crack instabilities leads to constant steady-state crack speeds equivalent to crack speeds in a linear solid with elastic modulus equal to the effective elastic modulus of his model solid. To conclude this Part, Thompson and Lewis use density functional theory to conduct a detailed study of the

Springer Proceedings in Physics, Volume 123
Computer Simulation Studies in Condensed-Matter Physics XIX
Eds.: D.P. Landau, S.P. Lewis and H.-B. Schüttler

surface structure of TiO_2 (110), making contact with a recent high-precision experiment. They find that bond angles converge relatively slowly with respect to model approximations, whereas bond lengths converge rapidly. Their results suggest changing the way surface structures for covalently bonded solids are reported, emphasizing bond lengths and angles, and not absolute positions.

In Part II, new models, methods, and perspectives are highlighted. In the first invited presentation Trebst et al. discuss the use of extended ensemble Monte Carlo methods to study equilibrium behavior in many-particle systems. They show how one possible method of optimization may be achieved by biasing the random walk by the inverse square root of the diffusivity. This approach is applied to a dense fluid, and they also show how to optimize parallel tempering simulations. Next, Ito describes the "Avogadro challenge" in which new computer technology in the form of Petaflop computing will enable us to approach simulation of macroscopic properties using microscopic models. The emphasis is on transport phenomena and both numerical estimates for properties as well as structures, e.g. vortices and bubbles, that result from such simulations are presented. Adler et al. then describe the visualization of nanostructures of carbon using AViz, a visualization package developed by the Computational Physics group at the Technion. The thrust of the discussion is on the challenges of using visualization to differentiate similar, competing structures. In the next paper Aoki describes methods for properly simulating anisotropic systems with periodic boundary conditions. A factor is introduced into the appropriate Lagrangian with the consequence that the normal pressure and the surface tension are given correctly. Sample data are shown for a system of soft spherocylinders. To conclude this part, De Raedt et al. introduce a model that is strictly causal and local. Although there are no principles of quantum mechanics involved, the model can correctly reproduce the single-spin expectation values and two-spin correlation functions in an Einstein-Podolsky-Rosen-Bohm experiment.

Part III contains four papers that examine non-equilibrium and dynamic behavior. O'Malley et al. have applied a dynamical Monte Carlo simulation approach to study a two-dimensional two-species lattice model describing the spread of an invading advantaged gene, allele or species into a disadvantaged resident population. Their results for the invasion front characteristics exhibit important qualitative differences from continuum-limit mean-field behavior. The differences are attributed to the absence of lattice discreteness and noise in the continuum-limit mean-field approximation. Next, Ogushi, Yukawa and Ito have performed molecular dynamics simulations for the dynamics and thermal structure of the gas-liquid interface in a three-dimensional Lennard-Jones system. They study the short- and long-time dynamical behavior of the density and temperature profiles which arise when a temperature gradient, imposed on an initial equilibrium state of the system, forces the nucleation of a gas-liquid interface. Their results demonstrate the existence of a temperature gap between the two emerging phases, indicative of an enhanced heat flow resistance across the interface. Kamimura, Yukawa and Ito have investigated the chemical reaction dynamics in far-from-equilibrium open systems within the framework of a stochastic model for a network of mutually catalytic replication reactions, controlled by an externally imposed energy flow. Their reaction-rate results show Arrhenius or non-Arrhenius behavior, respectively, under either low- or high-energy flow conditions. The authors suggest a possible alternative mechanism for biological systems to attain higher energy states in far-from-equilibrium situations. To close this part Hamad, Robb and Rikvold describe a comparative study of the dynamics of first- and second-order phase transitions by means of first-order reversal curve (FORC) analyses, applied to kinetic Monte Carlo simulation results for electrochemical deposition models. In the first-order case, their FORC results are indicative of a competition between the applied time-dependent electrochemical potential and the tendency towards phase ordering. In the second order case, by contrast, their results are characteristic of relaxational behavior towards a single equilibrium state.

Magnetic systems are featured in Part IV, beginning with Costa et al. who combine Monte Carlo and molecular dynamics simulations to study the microphysics of friction induced by magnetic forces for a model of a magnetic reading head moving across a magnetic film. They find that heat dissipation in this model system correlates to the appearance of magnetic vortices. Next, Machida, Iitaka, and Miyashita present a new technique for calculating linear response functions for quantum spin Hamiltonians on a large Hilbert space. It is based on a Chebyshev expansion of thermal and time-evolution operators that circumvents the need for high-dimensional diagonalization to obtain spin-spin correlation functions. Applied to a molecular nanomagnet, this technique gives results in close agreement with experiment. To conclude this section, Yoshioka et al. simulate cluster-cluster aggregation for a model system composed of Ising dipolar particles using molecular dynamics. They find that this system exhibits the behavior of attraction-limited cluster aggregation of dipolar particles. Various aspects of the dynamical properties of this model system are presented.

The last Part of this volume contains four papers on biological and soft condensed matter. Opening this Part is Shimada who reports on neuronal network simulations, aimed at elucidating the multiple internal states of neurons, and the transitions between such states that can be induced by cellular current injection, as recently observed in hippocampal slices from the brains of rats. The simulations demonstrate how multistable states can be induced in a network consisting of connected neurons that do not exhibit multistability in isolation. The comparison of different network topologies shows that shorter-range clustered network connectivity is required for the emergence of well-controlled network activity involving transitions between these states. Then, Laradji and Kumar have studied the dynamics of self-assembled multi-component lipid membranes using dissipative particle dynamics simulations. The effects on the dynamics of domain interface and surface tension, as well as lipid volume fraction, and trans-bilayer lipid distribution asymmetry, are investigated. Symmetric bilayers are found to exhibit a rich repertoire of dynamical behaviors, involving the coalescence and budding of local domain structures. Transbilayer asymmetry strongly affects both the curvature of the membranes and their dynamics. Next, Deserno gives a broad overview of solvent-free lipid bilayer simulations, their impact on our current understanding of the physics of these systems, and their relevance to biology. He outlines how the "solvent-free" modeling approach allows one to circumvent the explicit inclusion of the solvent by means of appropriately constructed effective pair potentials acting directly between the lipid monomers. A wide range of physically and biologically relevant properties can then be investigated, ranging from the bilayer self-assembly to their structure, equilibrium phase diagram and dynamics. Lastly, Ricci et al. report on Monte Carlo studies of the effects of boundaries on spatially confined two-dimensional colloidal crystals, based on a point particle model with inverse power law interaction potentials. They demonstrate that "flat" confining walls, will destroy positional long-range order in confined geometries, while locally enhancing orientational long-range order. Certain corrugated boundaries can restore positional long-range order, thereby illustrating the strong boundary sensitivity of such systems even for very distant boundaries.

Part I

Materials Properties

2

Accelerated Molecular-Dynamics Simulation of Thin Film Growth

K.A. Fichthorn and R.A. Miron

The Pennsylvania State University, University Park, PA 16802, USA

Abstract. Rare-event simulations can be limited by the wide range of time scales they must often probe. Using accelerated molecular dynamics in the framework of hyperdynamics, we overcome this limitation by identifying and consolidating groups of shallow minima into coarse states [1]. The method ensures a correct equilibrium sampling of the fast processes while enabling the simulation to be run on the time scale of the slow events. We demonstrate the method in accelerated molecular-dynamics simulations of the diffusion of Co clusters on Cu(001) [2].

2.1 Introduction

A significant challenge in simulating structural evolution in materials is to simulate long time and large length scales while accurately including atomic detail. Molecular-dynamics (MD) simulations can provide accurate details at the atomic scale. However, MD is not practical for simulating times or distances much beyond the nano scale. In many materials, dynamical evolution occurs through a series of "rare events", in which the system spends a long-time period in one potential-energy minimum before escaping and moving on to another. Transition-state theory (TST) can describe the average escape time [3]. Since the localized motion in the minima is not significant, dynamical evolution can be simulated as a series of long-time TST jumps between potential-energy minima. This is the aim of kinetic Monte Carlo (KMC) simulations. In principle, if a KMC simulation can incorporate all potential-energy minima of a system and the TST rates of all possible long-time jumps between the minima, then this technique can reach macroscopic times and larger length scales than MD, while retaining the accuracy of MD. In practice, however, there are still impediments limiting the wide applicability of KMC.

In many condensed-matter systems (e.g., amorphous solids, such as glassy polymers, ice, silicon, metals, as well as heteroepitaxial thin films), it is difficult and computationally expensive to classify all the potential-energy minima and the jumps between them. Hence, KMC methods have not been rigorously

Springer Proceedings in Physics, Volume 123
Computer Simulation Studies in Condensed-Matter Physics XIX
Eds.: D.P. Landau, S.P. Lewis and H.-B. Schüttler

applied to these systems and accurate prediction of their structural evolution remains a challenge. As we shall discuss below, recent innovations in accelerated MD simulations and efficient TST search algorithms have ameliorated some of these difficulties. These methods can probe time scales that are many orders of magnitude longer than those in conventional MD simulations while retaining the accuracy of conventional MD. However, the applicability of these and other related methods has an inherent limitation: The achievable time scale is limited by the fastest rate processes. In many systems, the rates of the various available processes can span many orders of magnitude. Particularly problematic is the case where fast processes are grouped together in pools of shallow energy minima that are separated from the rest of the phase space by high energy barriers. The system may perform millions of repeated transitions between these states before escaping away and, as a result, the overall progress of the simulation is curtailed significantly. This "small-barrier" problem has been recognized in studies of metal thin-film epitaxy [4, 5]. However, small barriers are ubiquitous and this problem could extend to a variety of rare-event systems, including protein folding, chemical reaction dynamics and catalysis at surfaces, polymer and glass dynamics, as well as transport on and within solids. We have recently developed a state-coarsening accelerated MD method [1] that could be beneficial for systems such as these and below, we discuss this method and its application to the growth of Co on Cu(001) [1,2].

2.2 Accelerated Molecular Dynamics

The basis for enhancing MD simulations is TST [3], which provides the escape rate $k^{\mathrm{TST}}_{A\to B}$ from a minimum A to a neighboring minimum B as a canonical average of the rate at which a trajectory beginning in state A will reach the dividing surface between A and B, i.e.,

$$k^{\mathrm{TST}}_{A\to B} = \frac{|v_\perp|}{2} \frac{\int_{\mathrm{A}} \delta^{\dagger}_{\mathrm{AB}}\ e^{-\beta V(\boldsymbol{x})} d\boldsymbol{x}}{\int_{\mathrm{A}} e^{-\beta V(\boldsymbol{x})} d\boldsymbol{x}} . \tag{2.1}$$

Here $\beta = 1/k_B T$, V is the potential energy, $\delta^{\dagger}_{\mathrm{AB}}$ is the delta function defining the location of the dividing hypersurface, and $v_\perp$ is the velocity component orthogonal to the dividing hypersurface. The integrals in (2.1) are evaluated over the domain of state A. MD simulations yield TST rates naturally, as the reciprocal of the canonical-average escape time achieved by integrating Newton's equations of motion. However, the escape time can far exceed the time scale that can be reached by MD at many temperatures of interest, making MD simulations impractical.

The first accelerated MD simulations [6–8] were based on importance sampling, recognizing that the ensemble average of (2.1) can be written as

$$k^{\mathrm{TST}}_{A\to B} = \frac{|v_\perp|}{2} \frac{\int_{\mathrm{A}} \delta^{\dagger}_{\mathrm{AB}}\ W(\boldsymbol{x})\ e^{-\beta V(\boldsymbol{x})} d\boldsymbol{x}/W(\boldsymbol{x})}{\int_{\mathrm{A}} W(\boldsymbol{x})\ e^{-\beta V(\boldsymbol{x})} d\boldsymbol{x}/W(\boldsymbol{x})} . \tag{2.2}$$

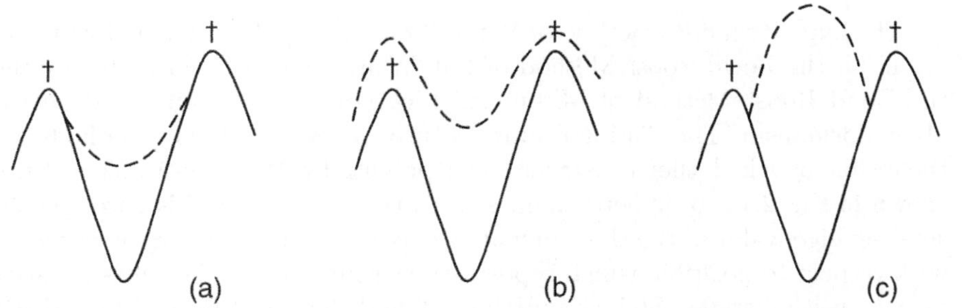

Fig. 2.1. Properties of the bias potential ΔV. The original potential V is shown by a solid line, $V'(= V + \Delta V)$ is shown by a dashed line, and transition states are indicated by †. In (**a**), an ideal bias potential is shown, which is higher than V near the minimum and becomes equal to V near transition states. The bias potential in (**b**) does not lead to $\Delta V = 0$ at the transition states, so detailed balance is not upheld. In (**c**), V' at the minimum is higher than the potential at the transition states, leading to insufficient sampling of the minimum

Here, W is a weighting function that can be defined as

$$W(\boldsymbol{x}) = exp(-\beta \Delta V), \tag{2.3}$$

where

$$\Delta V(\boldsymbol{x}) = V'(\boldsymbol{x}) - V(\boldsymbol{x}). \tag{2.4}$$

Substituting W of the form given by (2.3) into (2.2), the canonical average yielding k^{TST} can be written as

$$k^{\mathrm{TST}}_{A\to B} = \frac{|v_\perp|}{2} \frac{\langle \frac{\delta^\dagger_{\mathrm{AB}}}{W(\boldsymbol{x})} \rangle_{V'}}{\langle \frac{1}{W(\boldsymbol{x})} \rangle_{V'}}, \tag{2.5}$$

where $\langle \rangle_{V'}$ represents a canonical average on V'.

The bias potential ΔV should have several important properties, which are illustrated in Fig. 2.1. First, to accelerate the dynamics we would like to have $\Delta V \geq 0$ so that the depths of the potential minima are decreased. To uphold a detailed balance between various possible escape paths from a given minimum, V' must equal the original potential V at the transition states, or $\Delta V^\dagger = 0$. The bias potential should not otherwise corrupt the dynamics. In a properly constructed bias potential, the dynamics on V' over a time $\Delta t_{V'}$ is equivalent to the physical dynamics on V over a longer time Δt:

$$\Delta t = \Delta t_{V'} \langle e^{\beta \Delta V} \rangle_{V'}. \tag{2.6}$$

The time acceleration (or the boost) $\Delta t/\Delta t_{V'}$ increases with $\Delta V(\boldsymbol{x})$ and decreasing temperature. Since the acceleration is exponential in ΔV, very large accelerations are possible.

The hyperdynamics method of Voter [7,8], its simplified version by Steiner et al. [9], the Local Boost Method of Pal, Wang, and Fichthorn [10, 11], and the Bond Boost Method of Miron and Fichthorn [12] are all based on the above formalism. The distinguishing features of these different methods are the means by which they construct a bias potential with the desirable features shown in Fig. 2.1a. In hyperdynamics [7, 8], the bias potential is based on the smallest eigenvalue of the Hessian matrix of second derivatives of the potential with respect to position, which is positive near minima and becomes negative near transition states. Steiner's method [9] and the Local Boost Method [10, 11] are based on the values of single-particle energies, which exceed a preset threshold at transition states. As we will discuss in more detail below, the bias potential in the Bond Boost Method is based on the stretching of nearest-neighbor bonds past a threshold, at which transition-state crossing is assumed to occur [12].

Other approaches to accelerating MD simulations are based on temperature acceleration [5, 13] and using parallel simulations of many replicas of a system to enhance the chance of observing a rare event, in Voter's parallel replica dynamics [14]. In a related method, Henkelman and Jónsson use the Dimer Method [15–17] for finding transition states and constructing (harmonic) TST rate constants "on the fly" in a KMC algorithm without predefined lists of events used in traditional KMC. A review of these methods can be found in [18].

A problem with all of the above methods is that the achievable acceleration is limited by the event(s) with the lowest energy barrier(s). Particularly problematic is the case when pools of shallow states are present, as we illustrate in Fig. 2.2. The system may cycle many times between states $A - E$ in Fig. 2.2 before escaping over the high energy barriers to other states. In recent work, we proposed a method for addressing the small-barrier problem with accelerated MD. Our method is based on the Bond-Boost Method [12], which is a variant of hyperdynamics [7, 8]. We developed an extension of the Bond Boost Method to detect, on the fly, groups of recurrent shallow states and modify the potential-energy surface (PES) locally to consolidate them into large, coarse states. This procedure enables successful application of accelerated MD to systems exhibiting multiple-time scale processes and allows for substantially larger time boosts than can be achieved by accelerated MD alone.

In the Bond Boost Method, $\Delta V(\boldsymbol{x})$ [cf., (2.4)] is a function of the nearest-neighbor bond lengths $\{r_i\}$. It has a maximum value $\Delta V^{\max}$ at the local minimum configuration and goes to zero when the relative stretch or compression $\epsilon_i = (r_i - r_i^0)/r_i^0$ of any bond surpasses a threshold q, where r_i^0 is a local equilibrium bond length. The functional form is

$$\Delta V(\boldsymbol{x}) \equiv \Delta V^{\max} A(\epsilon^{\max}) \sum_{i=1}^{N_b} \delta V(\epsilon_i)\,, \tag{2.7}$$

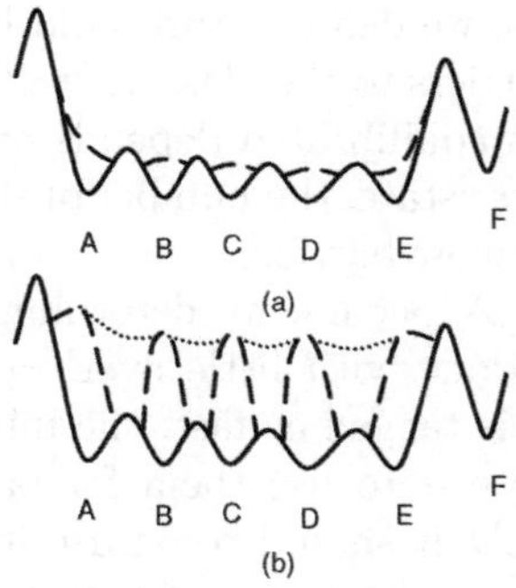

Fig. 2.2. One-dimensional illustration of the small-barrier problem. With an appropriate bias potential, as shown in (**a**), the system will be trapped in shallow states without a large time acceleration. Our solution is to combine a large boost with a bridging potential, as shown in (**b**), to consolidate states $A - E$ into one large minimum

where $\epsilon^{\max} = \max_i\{|\epsilon_i|\}$, N_b is the number of bonds included in the boost, $\delta V(\epsilon_i)$ is the boost applied to each bond, and $A(\epsilon^{\max})$ is an envelope that has values between [0..1] and becomes zero when $\epsilon^{\max} > q$. Each time the system reaches a new state, conjugate-gradient minimization is employed to find the new configuration $\{r_i^0\}$. Subsequently, the system is equilibrated in the new minimum and the boost potential is turned on. After a threshold q has been exceeded, minimization recommences to detect a new state after a waiting time has passed, and the cycle repeats. In the Bond Boost Method, the time boost is controlled by the magnitude of the boost potential $\Delta V^{\max}$ [cf., (2.7)], which can, in principle, be tuned. However, there is an upper limit on the achievable boost, as shown in Fig. 2.1c. With a small boost, appropriate for shallow minima [cf., Fig. 2.2a], the system will rapidly and repeatedly cycle between $A - E$ and escape to state F only over a much longer time. Since MD simulations are limited in the total number of time steps they can cover, the bulk of the simulation time is spent on these repeated transitions and evolution of the system is limited. Our proposed solution to this problem is to combine a very large boost, such as that shown in Fig. 2.1c, with "bridge potentials" ΔV^{bridge}, which span the transition states between states A, B, C, D, and E, as shown in Fig. 2.2b. In doing this, we consolidate the shallow states $A - E$ into a single, coarse state. This procedure is based on the assumption that equilibrium between the shallow states is reached long before any slow event $E \rightarrow F$ occurs. Since the "fast" dynamics become irrelevant on the time scale of the slow escape, we drop the requirement that $\Delta V(\boldsymbol{x}) = 0$ at the "fast" transition states. Thus, for the escape rate $k_{E\rightarrow F}$ the entire set $A \cup B \cup C \cup D \cup E$ acts as the "initial state" for the TST rate process.

The first issue in implementing the scheme shown in Fig. 2.2b is definition and detection of fast, recurrent processes. In a preliminary analysis, one can determine the range of energy barriers corresponding to the "slow" and "fast" time scales. Based on this, we define a threshold barrier ΔE^{th} and assign all processes having lower barriers to the class of "fast" processes. The validity of treating the fast states as equilibrated depends on the number of fast states consolidated into one coarse state, the number of slow exits, and on the energy gap between the fast and slow barriers.

When an event $M \to N$ occurs, we determine if the barrier $\Delta E^{\dagger}_{M\to N}$ is less than ΔE^{th}. This is done with little overhead using the Step-and-Slide method [19]. If a process is tagged as fast, the initial and final states M and N are stored. The purpose is to use them for pattern matching, i.e., when the system revisits state M it should recognize its transition to N as a fast process and activate the appropriate bridge potential. The initial and final states M and N for each process are defined locally and comprise only the pattern of bonds $\{r^0\}$ for the nearest neighbors $nn(M \to N)$ of the atom(s) that move in the transition $M \to N$: $M \equiv \{r^0_{nn(M\to N)}\}$. This formulation decouples local states from the global state and results in a linear increase of the storage requirement with system size.

The second element of our method is to construct a bridge potential between M and N. To this end, we define a potential $\Delta V^{\text{bridge}}_{\text{MN}}(\boldsymbol{x})$ that depends on the position(s) $\boldsymbol{x}_i$ of the moving atom(s) i along the transition path $M \to N$. $\Delta V^{\text{bridge}}_{\text{MN}}(\boldsymbol{x})$ is constructed as in (2.7), with $A(\epsilon^{\max})$ replaced by an envelope $A^{\text{bridge}}_{\text{MN}}(\boldsymbol{x}_i)$:

$$\Delta V^{\text{bridge}}_{\text{MN}}(\boldsymbol{x}) \equiv \Delta V^{\max} A^{\text{bridge}}_{\text{MN}}(\boldsymbol{x}_i) \sum_{i=1}^{N_b} \delta V(\epsilon_i) \,. \tag{2.8}$$

We define a sequence of p images $\{\boldsymbol{x}_i^0, .., \boldsymbol{x}_i^p\}$ along the approximate minimum-energy path (MEP) for the $M \to N$ process, so that the endpoints $\boldsymbol{x}_i^0$ and $\boldsymbol{x}_i^p$ coincide with the respective local minima. Generally, the MEP should be exactly calculated using a chain-of-states method such as the nudged elastic band [20]. The bridge envelope $A^{\text{bridge}}_{\text{MN}}(\boldsymbol{x}_i)$ is comprised of local contributions $A_l(\boldsymbol{x}_i)$ centered around each image $\boldsymbol{x}_i^l$:

$$A_l(\boldsymbol{x}_i) \equiv \max\left\{0, \alpha_l \left(1 - \frac{(\delta \boldsymbol{x}_i^l)^2}{w^2}\right)\right\} , \tag{2.9}$$

where $\delta \boldsymbol{x}_i^l = \boldsymbol{x}_i - \boldsymbol{x}_i^l$, and w and $\alpha_l \leq 1$ are parameters that control the shape of the bridge potential. Figure 2.3 illustrates properties of an ideal bridging potential. If all $\alpha_l = 1$, the PES is uniformly shifted and all barrier heights are preserved, as shown in Fig. 2.3a. However, it is desirable to smoothen the PES inside $A \cup B \cup C \cup D \cup E$ to promote efficient equilibrium sampling, as shown in Fig. 2.3b. Based on the potential energies of the images x_i^l, the parameters α_l can be chosen so that the modified PES is approximately flat. The parameter w controls the width of the bridge potential in the directions orthogonal to the MEP. In the case of an fcc crystal, w can be taken to be about half the nearest-neighbor distance.

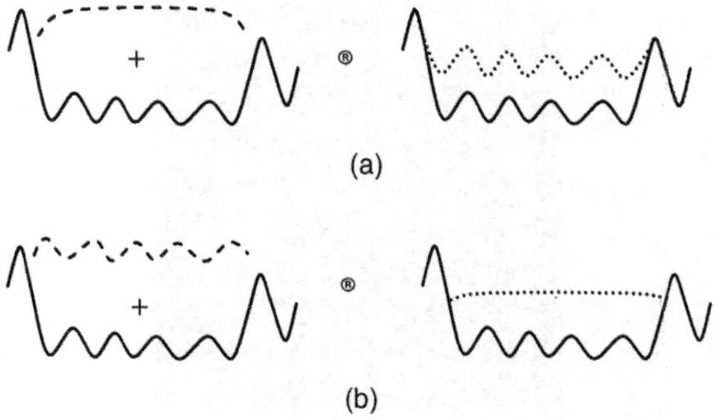

Fig. 2.3. One-dimensional illustration of properties of the ideal bridging potential. With a flat bridging potential, shown as the dashed curve to the left in (**a**), the structure of the original PES will be preserved. The ideal bridging potential, shown as the dashed curve to the left in (**b**), flattens the original curvature

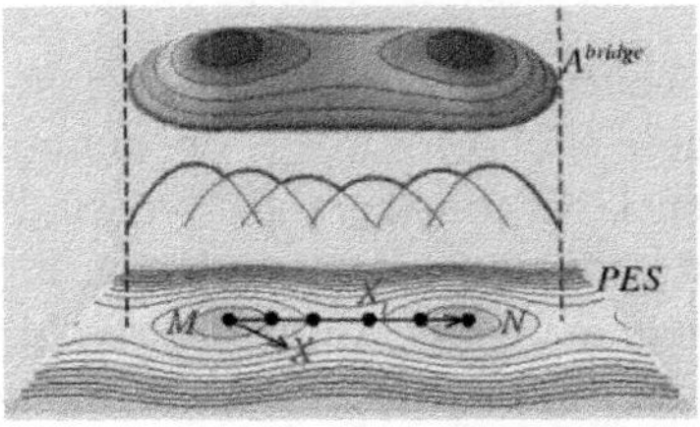

Fig. 2.4. Schematic shape of the bridge potential connecting two shallow minima $M-N$. $A_{\mathrm{MN}}^{\mathrm{bridge}}$ is a function of the distances from the instantaneous positions $\boldsymbol{x}_i$ to several images $\boldsymbol{x}_i^l$ placed along the MEP

The bridge envelope $A_{\mathrm{MN}}^{\mathrm{bridge}}(\boldsymbol{x}_i)$ is defined as

$$A_{\mathrm{MN}}^{\mathrm{bridge}}(\boldsymbol{x}_i) = \max_l \{A_l(\boldsymbol{x}_i)\}, \quad l = [0..p]. \tag{2.10}$$

The complete boost $\Delta V(\boldsymbol{x})$ is obtained by taking the envelope of all bond and bridge terms that are active at the particular instantaneous configuration, by merging (2.7) and (2.8):

$$\Delta V(\boldsymbol{x}) = \Delta V^{\max} \min_i \left(\max \left\{ A(\epsilon_i), A_i^{\mathrm{bridge}} \right\} \right) \sum_{i=1}^{N_b} \delta V(\epsilon_i). \tag{2.11}$$

A diagram illustrating the main elements of the boost and bridging potential is shown in Fig. 2.4.

The complete algorithm is as follows. After each event, we start off with a low boost, i.e., $\Delta V^{\max} \lesssim \Delta E^{th}$, that preserves the correct fast dynamics.

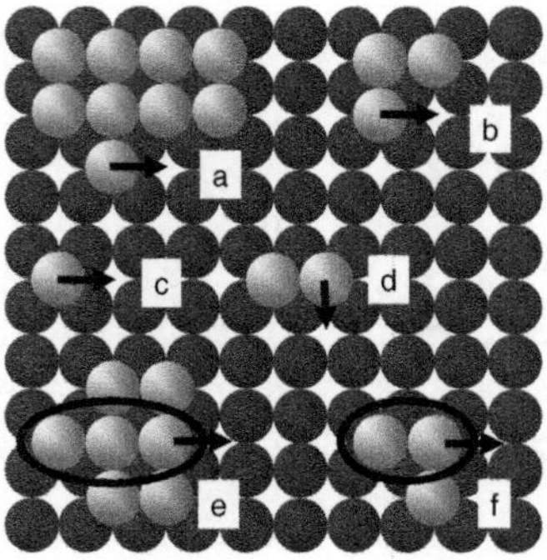

Fig. 2.5. Rate processes involved in the growth of Co (shown as light atoms) on Cu(001) (dark atoms). Processes include atom hopping along island edges (**a**), trimer rotation (**b**), adatom hopping (**c**), dimer hopping (**d**), heptamer hopping via trimer shearing (e), and trimer hopping via dimer shearing (**f**)

The low boost allows accurate dynamics of the fast events, albeit with low acceleration. As each new shallow state is encountered, it is stored and the appropriate bridge potential is constructed. Eventually, the shallow states are exhausted. If no new state is encountered during a predefined waiting time $t^{\rm wait}$, the boost strength is increased to the desired high value, the bridges are activated, and the simulation switches to the time scale of the slow events. We choose $t^{\rm wait}$ larger than the average waiting time for a process having the threshold barrier, i.e., $t^{\rm wait} > \nu^{-1}\exp(\beta\Delta E^{th})$, where ν is an appropriate prefactor. In each new state, the simulation code performs local pattern matching against stored states to find the applicable bridge potential terms. With an efficient implementation of the state recognition algorithms, the overhead incurred by this method is in general less than 10% of the normal simulation time for our test cases, and should remain low as long as the number of fast processes is finite.

To illustrate our method, we simulated kinetic phenomena related to the heteroepitaxial thin-film growth of Co on a Cu(001) surface [1,2]. The interaction potential is based on the Tight-Binding Second-Moment Approximation, as discussed by Miron and Fichthorn [2]. Some of the processes involved in growth are shown in Fig. 2.5. This system exemplifies the small-barrier problem. For example, hopping of atoms along island edges (Fig. 2.5a) is about 10^6 faster than adatom hopping (Fig. 2.5c) at room temperature and, once a trimer is formed, its rotation, shown in Fig. 2.5b, is about 10^8 times faster than adatom hopping. Rapid trimer rotation leads to a pool of 24 shallow states. Without bridge boosting, even accelerated MD simulations would be hopelessly limited by the short timescale of edge hops.

We implement bridge boosting using a threshold of $\Delta E^{th} = 0.4\,$eV. For constructing the bridging potentials, we approximate the MEP as a straight line, which yields a particularly simple implementation of (2.9) and (2.10). After each event, a low boost of $\Delta V^{\max} = 0.2\,$eV is maintained until no new

Table 2.1. Energy barriers ΔE for various diffusion processes of Co/Cu(001). Static values are obtained with the Step-and-Slide method [19]

Process	ΔE^{static} (eV)	ΔE^{MD} (eV)
Adatom hop	0.63	0.63
Dimer hop	0.62	0.63
Adatom edge hop	0.30	n.a.
Trimer rotation	0.10	n.a.
Trimer hop	0.64	0.65
Heptamer hop	0.56	0.57

process occurs for a time $t^{\text{wait}} = 10^{-2}\exp(\beta\Delta E^{th})$ ps. Then, the boost is increased to $\Delta V^{\text{max}} = 0.6$ eV. By overcoming the small-barrier limitation, we achieve boosts [cf., the discussion pertaining to (2.6)] ranging from 10^4 at 450 K to 10^8 at 250 K and slow island diffusion is correctly captured. Hopping along island edges is much faster than events that lead to center-of-mass motion of small islands, which are mainly collective shearing mechanisms. Dimers, trimers and heptamers have a high mobility, comparable to that of the isolated adatom. The trimer hops via concerted jump of two atoms, as shown in Fig. 2.5f while the heptamer hops via concerted shearing of three atoms in the middle row (cf., Fig. 2.5e). A less favorable mechanism for heptamer hopping involves an edge adatom climbing on top and descending on the other side of the compact 6-atom island formed by the remaining atoms. Single adatom exchange can occur in this system, however due to the high energy barrier (0.92 eV) it is not active on the time scale and temperature range covered in these simulations. Energy barriers obtained from static calculations and bridged, accelerated MD simulations are shown in Table 2.1, where we find excellent agreement between the two.

Thus, we developed a general method for allowing accelerated MD simulations to cover multiple-time scale processes for rare-event dynamics. By assuming that a separation of fast and slow time scales exists, the method treats fast processes as equilibrated on the slow time scale, and consolidates pools of shallow minima detected on-the-fly into coarser states. The slow dynamics is preserved and the simulation time scale can be tuned to the slow events of interest. Above, we illustrated the application of this method to the diffusion of Co clusters on Cu(001). In other studies, we applied this method to investigate the growth of Co on Cu(001) [2]. The method could have wide application in materials simulation of rare-event dynamics.

References

1. R.A. Miron, K.A. Fichthorn, Phys. Rev. Lett. **93**, 138201 (2004).
2. R.A. Miron, K.A. Fichthorn, Phys. Rev. B 72, 115433 (2005).

3. P. Hänggi, P. Talkner, M. Borkovec, Rev. Mod. Phys. **62**, 251 (1990).
4. O. Biham, I. Furman, M. Karimi, G. Vidali, R. Kennett, H. Zeng, Surf. Sci. **400**, 29 (1998).
5. M.R. Sørensen, A.F. Voter, J. Chem. Phys. **112**, 9599 (2000).
6. E.K. Grimmelmann, J.C. Tully, E. Helfand, J. Chem. Phys. **74**, 5300 (1981).
7. A.F. Voter, J. Chem. Phys. **106**, 11 (1997).
8. A.F. Voter, Phys. Rev. Lett. **78**, 3908 (1997).
9. M.M. Steiner, P.-A. Genilloud, J.W. Wilkins, Phys. Rev. **B 57**, 10236 (1998).
10. S. Pal, K.A. Fichthorn, Chemical Engineering Journal **74**, 77 (1999).
11. J.-C. Wang, S. Pal, K.A. Fichthorn, Phys. Rev. **B 63**, 85403 (2001).
12. R.A. Miron, K.A. Fichthorn, J. Chem. Phys. **119**, 6210 (2003).
13. F. Montalenti, M.R. Sørensen, A.F. Voter, Phys. Rev. Lett. **87**, 126101 (2001).
14. A. Voter, Phys. Rev. B **57**, R13985 (1998).
15. G. Henkelman, H. Jónsson, J. Chem. Phys. **111**, 7010 (1999).
16. G. Henkelman, H. Jónsson, Mat. Res. Soc. Symp. Proc. **677**, AA8.1.1 (2001).
17. G. Henkelman, H. Jónsson J. Chem. Phys. **115**, 9657 (2001).
18. A.F. Voter, F. Montalenti, T.C. Germann, Annu. Rev. Mater. Res. **32**, 321 (2002).
19. R.A. Miron, K.A. Fichthorn, J. Chem. Phys. **115**, 8742 (2001).
20. G. Henkelman, B.P. Uberuaga, H. Jónsson", J. Chem. Phys. **113**, 9901 (2000).

3

Simulating the Interaction of High Intensity Optical Pulses with Nanostructured Optical Devices

W.M. Dennis and C.M. Liebig

The University of Georgia, Athens, GA 30602, USA

Abstract. Due to nonlinear optical effects the reflective properties of dielectric interference filters may be modified at high intensities. In this work we discuss three dielectric filters: a quarter-wave dielectric stack, a continuously varying rugate filter, and a rugate filter constructed of two discrete index materials. The finite difference time domain (FDTD) method was used to simulate the reflectance of both low and high intensity pulses from these optical devices. At high intensities a decrease of the filter reflectance as well as distortion of the reflected pulse was observed.

3.1 Introduction

Over the past decade it has become possible to purchase table-top laser systems that are capable of generating ultrashort optical pulses with sufficiently high intensity to induce nonlinear optical effects in many of the materials that are used in the construction of optical components. Dielectric structures that have been well characterized as high reflectors at low intensities are expected to change their reflective properties when interacting with high-intensity pulses. To explore the effects high intensity pulses reflecting from optical filters we use the finite difference time domain (FDTD) method. FDTD is used to numerically integrate the Maxwell curl equations to obtain both the electric and magnetic fields [1]. While FDTD techniques were originally developed for isotropic media, rapid advancements in the speed, memory and storage of electronic computers has enabled FDTD methods to be applied to a wide range of problems including optical pulse-material interactions on the femtosecond time-scale [2].

Dielectric high-reflecting optical filters have been studied for over a century and yet still find extensive use in modern optics laboratories. Due to their high damage thresholds dielectric filters have found uses ranging from the narrow band high reflectors to cavity mirrors used in lasers [3–6]. The simplest quarter wave dielectric stack structures consist of a repeated sequence of $\lambda/4$ layers of two materials with different refractive indices. In order to both

Springer Proceedings in Physics, Volume 123
Computer Simulation Studies in Condensed-Matter Physics XIX
Eds.: D.P. Landau, S.P. Lewis and H.-B. Schüttler

optimize the reflected bandwidth and to eliminate the reflectance of higher-order harmonics, there have been several variations on the design of optical filters. These include gradient-index structures or rugate structures where the index of refraction is engineered to vary continuously in a sinusoidal manner along the direction of propagation [7, 8]. Alternatively, rugate structures can be constructed from a series of nanoscale discrete layers which produce the effect of a sinusoidal refractive index modulation over distances comparable with the wavelength of light [9]. In this paper we use the FDTD method to investigate changes in the reflection of high intensity ultrashort optical pulses from dielectric thin-film structures.

3.2 Multilayer Dielectric Structurs

The computer simulations described in this paper model the reflection of ultrashort pulses from three types of multilayer thin-film structures. The first structure is a simple quarter-wavelength dielectric stack that consists of a repeated sequence of $\lambda/4$ layers of two materials with different indices of refraction [10]. The quarter-wave stack modeled in our simulations was designed to have high reflectance at a central wavelength of 800 nm and was comprised of 14 cycles of Ta_2O_5 and SiO_2 on a SiO_2 substrate. The second structure studied is an idealized rugate structure in which the refractive index varies sinusoidally. The third structure is a practical rugate design that uses a repeating unit comprised of multiple nanoscale layers (of two differing materials) such that the spatially averaged refractive index varies sinusoidally on the length-scale of the optical wavelength. This latter approach has proven successful because rugate structures appear to be quite robust with respect to variations from a purely sinusoidal index profile and while not perfect, these discrete rugate filters have similar properties to the continuous rugate [9].

Figure 3.1 shows the structure of the quarter-wave dielectric stack and both rugate filters as well as their reflectance spectra.

3.3 Finite Difference Time Domain Simulations

The FDTD simulations were performed using an in-house 1-D finite difference time domain (FDTD) code implemented in Fortran 95. The FDTD method [1] uses the Yee algorithm [2] to directly integrate the Maxwell curl equations for the real-valued electric and magnetic fields. The Yee algorithm can be adapted to take into account both dispersive linear and nonlinear media. In our code we employ the auxiliary differential equation method which uses the inverse Fourier transform to convert the constitutive relation between $\boldsymbol{D}(\omega)$ and $\boldsymbol{E}(\omega)$ into a differential equation that can be discretized and integrated in tandem with the Maxwell curl equations. Linear dispersion can be introduced in terms of either Debye or Lorentz models [11]; third-order nonlinear optical

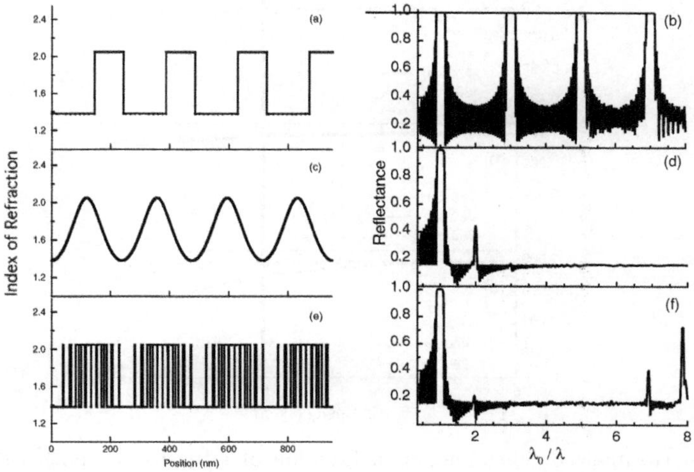

Fig. 3.1. Refractive index profiles for (**a**) four cycles of a quarter-wave dielectric stack interference filter, (**c**) four cycles of a continuous varying rugate filter and (**e**) four cyles of a 22 layer rugate filter constructed from Ta_2O_5 and SiO_2 layers. Panes (**b**), (**d**) and (**f**) show the corresponding reflectance spectra. The reflectivity of the rugate structures are suppressed at $n\lambda_0/\lambda$ as compared with the quarter-wave stack

behavior can be introduced as either instantaneous Kerr or delayed Raman processes [12].

The FDTD method is particularly appropriate for simulating multilayer structures as it enforces the field boundary conditions between dissimilar media. Furthermore, the fine grid required by the FDTD method is also necessary to accurately describe the rapidly spatially varying refractive index profile associated with the multilayer structures. A Kerr nonlinearity was included for Ta_2O_5 using an algorithm based upon the work of Goorjian et al. [12], however since the value of $\chi^{(3)}$ for SiO_2 is nearly two orders of magnitude lower than that of Ta_2O_5, nonlinear effects in the SiO_2 layers were neglected.

The simulations described in this paper were performed with a spatial resolution of at least $\lambda/100$ with some simulations being performed at $\lambda/200$ to test for convergence. In order to validate our code we used the results from low intensity FDTD simulations to generate reflectance spectra that could be compared with analytic reflectance spectra calculated using the characteristic matrix method as detailed below. The inclusion of nonlinear optical effects was tested by studying soliton propagation in a dispersive nonlinear medium and comparing our results to those of [12].

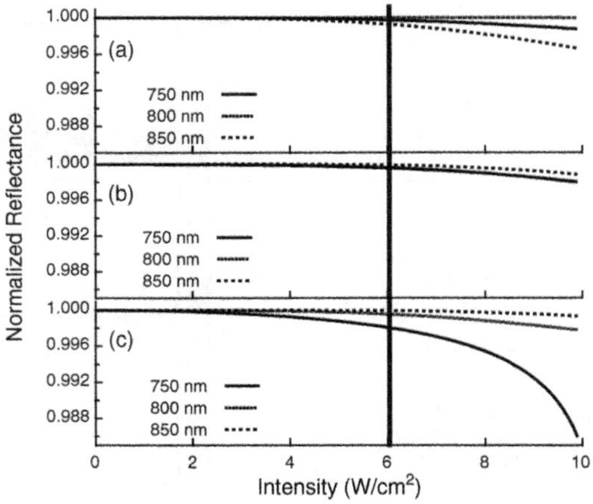

Fig. 3.2. Normalized reflectance as a function of intensity for both (**a**) quarter-wave dielectric stack filters, (**b**) continuous rugate, and (**c**) discrete rugate filters at wavelengths 750 nm, 800 nm and 850 nm. The vertical bar shows the intensity damage threshold for Ta_2O_5 for a 50 fs optical pulse at a wavelength of 800 nm [14]

3.4 Results and Discussion

Using the FDTD method we have simulated the reflection of 50 fs optical pulses from optical filters described in Sect. 3.2, for intensities between $\sim 10\,\mu\text{W/cm}^2$ and $\sim 10\,\text{TW/cm}^2$ at three wavelengths 750 nm, 800 nm and 850 nm. Figure 3.2 plots normalized reflectance as a function of incident pulse intensity for the quarter-wave stack and the rugate structures. Figure 3.2 illustrates that while high intensity pulses affect the performance of both structures they do so to different extents. At all three wavelengths studied, the reflectance of the quarter-wave dielectric stack changes only slightly as the intensity of the incident pulse increases, furthermore the reflectance of this device is not significantly reduced even well above the damage threshold. The rugate structure exhibits a more exaggerated behavior. At pulse wavelengths equal to or longer than the central wavelength of the structure, the reduction in reflectance, while somewhat greater than that observed for the quarter-wave stack, is still negligible. However, at shorter wavelengths the reflectance is noticeably reduced. The sensitivity of the rugate structure to incident pulses at wavelengths shorter than the central wavelength can be understood in terms of the following mechanism: Ta_2O_5 has a positive $\chi^{(3)}$ which causes the index of refraction to increase with increased intensity; the optical thickness of the Ta_2O_5 layers is therefore increased, effectively tuning the filter to longer wavelengths [13]. Similar effects have been observed in simulations of nonlinear Fabry-Perot cavities [3].

Finally, we point out that while the onset of nonlinear optical effects in the rugate structure occurs quite close to the damage threshold of the Ta_2O_5 layer, some of the more pronounced nonlinear optical effects (such as significant pulse shape distortion) that were only observed at intensities above the damage threshold of the Ta_2O_5/SiO_2 structure, will become important in structures which include layers of higher $\chi^{(3)}$ materials, e.g., Nb_2O_5 [4].

3.5 Conclusions

In this work we have used computer simulations based upon the FDTD method to investigate the interaction of high-intensity ultrashort optical pulses with two multilayer dielectric structures: a quarter-wave dielectric stack, continuous rugate filter and a discrete rugate filter. A simulation of the reflection of a low intensity, broad-band pulse was used to determine the reflectance spectra of both structures. The resulting spectra were found to be in good agreement with reflectance spectra calculated using analytical methods.

For computer simulations using 50 fs pulses, we observed that as intensity was increased, nonlinear optical effects caused higher optical harmonics to become evident in the reflected pulse spectra, while intensity dependent refractive index (n_2) effects caused changes in the optical thickness of the Ta_2O_5 layers, effectively detuning the optical properties of the structure. Our simulations confirm that the quarter-wave dielectric stack and the continuous rugate maintain their reflective properties up to extremely high intensities. However, the discrete rugate structures, while still quite robust, were more susceptible to nonlinear optical effects at high intensities, especially when used at wavelengths shorter than the center wavelength for which the rugate was designed.

References

1. A. Taflove, S.C. Hagness, *Thin-Film Optical Filters.* 2nd edn. (Artech House, Boston, 2000).
2. K.S. Yee, IEEE Trans. Antennas and Propagation **14**, 302 (1966).
3. S.A. Basinger, D.J. Brady, J. Opt. Soc. Am. B **8**, 1504 (1994).
4. T. Hashimoto, T Yoko, Appl. Opt **34**, 2941 (1995).
5. H. Szymanowski, O. Zabeida, J.E. Klemberg-Sapieha, L. Martinu, J. Vac. Sci. Technol. A **23**, 241 (2005).
6. X. Wang, H. Masumoto, Y. Someno, T. Hirai, J. Vac. Sci. Technol. A **17**, 206 (1999).
7. W.H. Southwell, J. Opt. Soc. Am. A **5**, 1558 (1988).
8. B.G. Bovard, Appl. Opt. **32**, 5427 (1993).
9. H.A. Macleod, *Thin-Film Optical Filters.* 3rd edn. (Institute of Physics Publishing, Bristol, 2001).
10. M. Born, E. Wolf, *Principles of Optics* 6th edn. (Pergamon, Oxford,1980).
11. R.M. Joseph, S.C. Hagness, A. Taflove, Opt. Lett. **16**, 1412 (1991).

12. P.M. Goorjian, A, Taflove, Appl. Opt. **17**, 180 (1992).
13. For example see: R.W. Boyd, *Nonlinear Optics.* 2nd edn. (Academic Press, Amsterdam, 2003)
14. J. Jasapara, A.V.V. Nampoothiri, W. Rudolph, Phys. Rev. B **63**, 1 (2001).

4

Crack Motion Revisited

F.F. Abraham

[1] Lawrence Livermore National Laboratory, Livermore CA
[2] University of Georgia, Athens GA

In our recent study of brittle fracture [1], we showed that hyperelasticity, the elasticity at large strain, plays a governing role in the onset of the crack instability from unidirectional motion. We discovered a simple, yet remarkable, scaling based on an *effective* elastic modulus for our modelled solid (the secant modulus at the stability limit) which led to successful predictions for the onset speed of the crack instability. We have now applied this scaling to the same modelled solid with the exception that the crack is constrained to travel unidirectional irrespective of its speed. This allows the crack to achieve a unique steady-state speed that has a dependence on hyperelasticity. Using our scaling law, we find that the steady-state crack speed scales to a constant value equal to a crack speed of a linear solid with our *effective* elastic modulus. We discuss how this finding is related to simple spring dynamics.

Our modelled solid is based on a generalized bilinear force law composed of two spring constants, one associated with small deformations (k_1, $r \leq r_{\rm on}$) and the other associated with large deformations ($k2$, $r > r_{\rm on}$). This is shown in Fig. 4.1. This has served as a useful model, allowing us to investigate the generic effects of hyperelasticity by changing the relative magnitude ($\bullet = k2/k1$) and transition distance ($r_{\rm on}$) of the potential. This bilinear force was used earlier to investigate the dynamics of cracks constrained to remain straight [2], and the results of a particular set of simulations are shown in Fig. 4.2a. In this study, we adopt a bilinear potential where $k_2 = 2k_1$ (a model material with elastic stiffening). We choose $r_{\rm break} = 1.17$. In the harmonic system ($r_{\rm break} = r_{\rm on}$), the crack achieves a propagation velocity around 90% of the Rayleigh wave speed. The simulations reveal that with a local stiffening zone around the crack tip, the crack propagates at super-Rayleigh velocities. Figure 4.2a plots the crack propagation velocity as a function of potential parameter $r_{\rm on}$. We observe that the earlier the hyperelastic effect occurs in the stretch, the larger the limiting velocity.

Depending on the choice of interatomic potential parameters for our model solid, we had defined an *effective* elastic modulus which has a remarkable relationship to the crack's instability dynamics (4.1). We discovered that the

Springer Proceedings in Physics, Volume 123
Computer Simulation Studies in Condensed-Matter Physics XIX
Eds.: D.P. Landau, S.P. Lewis and H.-B. Schüttler

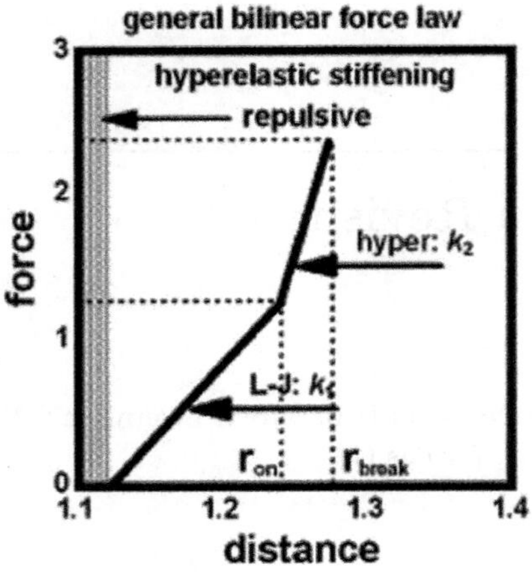

Fig. 4.1. The *effective* spring constant k_{eff} defined graphically for the bilinear force

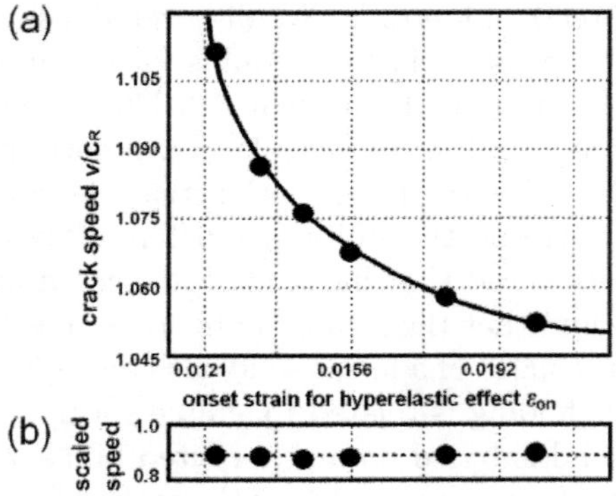

Fig. 4.2. (**a**) Constrained crack propagation speed as a function of parameter r_{on}. (**b**) Individual steady-state speeds scaled by the square-root of k_{eff})

crack's speed at its instability onset collapses onto a universal linear dependence with this new modulus. This finding allowed for successful predictions for the crack's instability speed in nonlinear materials. Figure 4.1 defines, graphically, our choice for an *effective* spring constant k_{eff} of the bilinear force law. It is simply the slope of the vector sum of the maximum piecewise linear forces defined by the bilinear force law as shown in Fig. 4.1,

$$k_{\mathrm{eff}} = k_1 \left[(r_{\mathrm{on}} - r_0) \,/\, (r_{\mathrm{break}} - r_0) \right] + k_2 \left[(r_{\mathrm{break}} - r_{\mathrm{on}}) \,/\, (r_{\mathrm{break}} - r_0) \right] \,. \quad (4.1)$$

It is simply related to the *secant modulus* bounded at the mechanical stability limit.

Plotting, in Fig. 4.2b, the scaled speeds (steady-state speed divided by the square-root of k_{eff}), we see a remarkable collapse of the data on to a horizontal straight line. Hence, for determining the steady-state speed of a constrained dynamic brittle crack, this finding allows one to model the bilinear material as a linear solid with the effective spring constant described by (4.1). This remarkable scaling has a very general validity.

References

1. F.F. Abraham, JMPS **53**, 1071 (2005).
2. M. Buehler, F.F. Abraham, H. Gao, Nature **426**, 141 (2003).

5

Deconstructing the Structural Convergence of the (110) Surface of Rutile TiO_2

S.J. Thompson and S.P. Lewis

University of Georgia, Athens GA 30602, USA

Abstract. In response to a recent high-precision LEED-IV experiment, we have carried out a detailed reexamination of the (110) surface structure of rutile TiO_2 using first-principles total-energy methods. Our previous analysis showed that bond lengths between surface atoms converge much faster with respect to model approximations than do relaxed absolute atomic positions. This difference in convergence rate is due to the relatively slow convergence of bond angles. We argue that this analysis favors altering the way structural data are typically reported for covalently bonded solids from absolute atomic positions to more physically meaningful quantities such as bond lengths and angles.

5.1 Introduction

TiO_2 is prominent in both applications and basic research, as well as a starting point for numerous first-principles investigations [1]. A recent quantitative low-energy electron diffraction (LEED-IV) experiment [2] has produced what is arguably the highest precision measurement to date of the structure of the (110) surface of rutile TiO_2. Comparison with previously published surface x-ray diffraction and first-principles theoretical investigations has revealed some significant discrepancies in the reported atomic positions that can be resolved if the system is described in terms of bond lengths in lieu of relaxed atomic positions [3].

Additionally, we have shown that bond lengths converge much faster with respect to the thickness of the slab used to model the surface than do relaxed atomic positions [3], which has been the typical method of reporting data for this and other covalently bonded solids. We will show that the difference in convergence rate can be explained by the relatively slow convergence of bond angles, which are generally 'softer' degrees of freedom than bond lengths. Since bond lengths and angles are physically meaningful quantities, and they clearly describe the geometry of the surface atoms without imposing an arbitrary reference point, we argue in favor of reporting bond lengths and bond angles as a more physically relevant description of the surface.

Springer Proceedings in Physics, Volume 123
Computer Simulation Studies in Condensed-Matter Physics XIX
Eds.: D.P. Landau, S.P. Lewis and H.-B. Schüttler

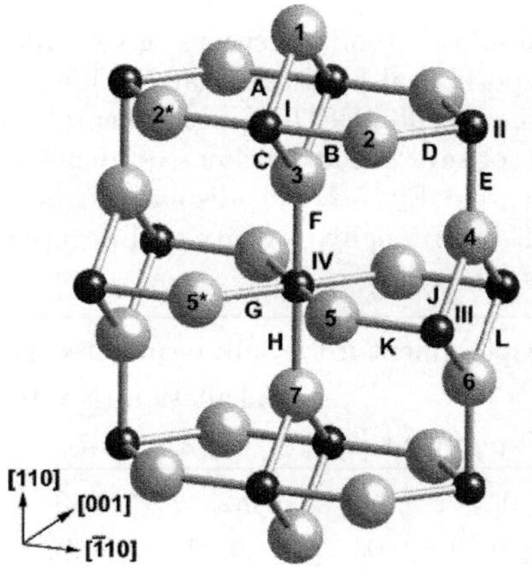

Fig. 5.1. Model of the TiO_2 (110) surface showing 3 trilayers. Atoms in the two trilayers nearest the surface are numbered using Arabic numerals for O atoms (*large, light spheres*) and Roman numerals for Ti atoms (*small, dark spheres*). An * denotes an atom paired via symmetry to the atom of the same number. The letters *A* through *L* (omitting I) label bonds

5.2 Method

We used the Vienna Ab-initio Simulation Package (VASP) [4, 5] to perform our ab-initio calculations, which have been carried out within the framework of density functional theory. Specifically, we used the generalized gradient approximation for exchange and correlation as well as ultrasoft pseudopotentials to model core-valence interactions for O and Ti atoms, where the semi-core 3p electrons of Ti were included in the valence manifold. The Kohn-Sham wave functions were expanded using plane waves up to a kinetic-energy cutoff of 396 eV, and the Brillouin zone was sampled using a $2 \times 4 \times 2$ Monkhorst-Pack grid. Detailed theoretical background for this methodology can be found in [6]. Additional computational details of these calculations can be found in [3].

In our surface calculations, we modeled the system as a finite slab of stacked TiO_2 trilayers periodically reproduced normal to the surface direction. Each TiO_2 trilayer comprises a central TiO plane with a sparse layer of so-called 'bridging' oxygen atoms symmetrically placed above and below the central plane. A three trilayer slab is shown in Fig. 5.1. A vacuum spacing of 19.6 Å, corresponding to the width of 6 trilayers, was used since it was found to be very well converged. The thickness of the slab, in terms of the number of trilayers, is an important convergence parameter for the calculations and is discussed in more detail below.

Table 5.1. Relaxed absolute atomic positions, given as displacements from bulk-like positions for a hypothetical bulk-terminated TiO_2 (110) surface. Results are presented from both the recent LEED-IV experimental study [2] and our calculations, which are shown for several slab thicknesses (denoted by number of trilayers). Atom labels are taken from Fig. 5.1, and displacements are in the [110] direction unless otherwise noted. Additionally, positive and negative directions are specified by the coordinate axes shown in Fig. 5.1

Atomic displacement from bulk terminated positions (Å)					
		Calculation (No. trilayers in slab) [3]			
Atom label	Experiment [2]	5	7	9	11
Ti (I)	0.25 ± 0.03	0.29	0.33	0.42	0.43
Ti (II)	-0.19 ± 0.03	−0.10	−0.07	−0.03	−0.03
O (1)	0.10 ± 0.05	0.09	0.13	0.23	0.23
O (2)	0.27 ± 0.08	0.24	0.27	0.32	0.32
O (2)[$\bar{1}$10]	0.17 ± 0.15	0.06	0.05	0.05	0.05
O (2*)[$\bar{1}$10]	-0.17 ± 0.15	−0.06	−0.05	−0.05	−0.05
O (3)	0.06 ± 0.10	0.07	0.11	0.19	0.19
Ti (III)	0.14 ± 0.05	0.17	0.23	0.32	0.32
Ti (IV)	-0.09 ± 0.07	−0.04	−0.03	0.00	0.00
O (4)	0.00 ± 0.08	0.06	0.09	0.13	0.13
O (5)	0.06 ± 0.12	0.06	0.09	0.15	0.15
O (5)[$\bar{1}$10]	0.07 ± 0.18	0.03	0.02	0.01	0.01
O (5*)[$\bar{1}$10]	-0.07 ± 0.18	−0.03	−0.02	−0.01	−0.01
O (6)	0.00 ± 0.17	0.03	0.07	0.10	0.11
O (7)	0.01 ± 0.13	0.02	0.04	0.10	0.09

5.3 Results

Table 5.1 shows our calculations of the relaxed absolute atomic positions for slabs containing 5 to 11 trilayers, where we have only investigated slabs with an odd number of trilayers due to odd-even oscillations [7]. As previously shown, surface energy does not converge until the slab thickness reaches approximately 13 trilayers [3, 7], whereas the displacements of the atoms from their bulk-terminated positions converge to within 0.01 Å for slabs with as few as 9 trilayers.

When comparing the results shown for the most recent experimental data [2] which is also shown in Table 5.1, one notices remarkably good agrement with the displacements that occur in our 5 trilayer slab, even though the remainder of the data in the table shows that these displacements are relatively far from the converged results. As we have previously demonstrated [3], this agreement occurs for two reasons. First, bond lengths converge much faster than atomic displacements; specifically, we found that, for a 5 trilayer slab,

Table 5.2. Angles formed when atomic positions are projected onto the (001) plane. Results are shown for both the recent LEED-IV experiment [2] and our calculations. The angles are denoted by $(\alpha - \beta - \gamma)$ where the angle is between the $(\alpha - \beta)$ and $(\beta - \gamma)$ bonds, and the atom labels (α, β, γ) are shown in Fig. 5.1

	Projected bond angles in degrees				
		Calculation (No. trilayers in slab) [3]			
Angle label	Experiment [2]	5	7	9	11
O(3)-Ti(I)-O(2)	90.5±2.9	88.6	88.1	87.0	87.0
O(2)-Ti(II)-O(4)	112.7±3.0	105.3	105.4	105.9	106.0
Ti(I)-O(2)-Ti(II)	156.7±4.9	166.0	166.5	167.1	167.0
O(6)-Ti(III)-O(5)	100.9±6.6	92.9	93.4	94.4	94.3
O(5)-Ti(IV)-O(7)	81.5±2.2	85.3	83.7	82.5	82.4
Ti(III)-O(5)-Ti(IV)	178.9±6.2	178.2	177.1	176.9	176.7

the calculated bond lengths are within one percent of their converged values. Additionally, the experimental values appear to be based on a structural model in which the third TiO layer down from the surface is in the bulk geometry. This is the also the case for our 5 trilayer slab, due to the mirror symmetry of the slab about the central TiO layer.

These results then lead us to question why significantly thicker slabs are required for atomic displacements to converge than for bond lengths. To answer this, we look at the convergence of bond angles with respect to slab thickness. It can be shown that 8 distinct bond angles exist for each trilayer. However, if one looks at the angles formed when the atoms are projected onto the (001) plane, there are only 3 distinct angles per trilayer. While the projected angles *are* related to the true bond angles, they are no longer a direct physical quantity. However, they end up giving a clearer picture of the reason that atomic displacements converge so much slower than bond lengths. The value of this projection approach lies in the fact that all atomic relaxations in the [001] direction are symmetry forbidden. Thus, the projected angles, shown in Table 5.2, clearly relate to the relative positions of the atoms along the surface [110] direction, which, as seen in Table 5.1, is the direction of slowest convergence.

As Table 5.2 shows, our calculated angles converge when the slab thickness reaches 9 trilayers, which is the same slab thickness for which the displacements of the atoms from their bulk-terminated positions converged. Interestingly, this consistent convergence demonstrates that our larger displacements in thicker slabs is a result of two different effects. First, Table 5.2 shows that as slab thickness increases, up to 9 trilayers, the projected angles for the surface layer change. Second, for thicker slabs, vertical-relaxation errors for deeper atoms compound the errors of the outer atoms. As an analogy, imagine a ruler in which the spacing between centimeter marks is actually only 0.9 cm.

The distance of the 1-cm mark from zero is only off by 0.1 cm, but the distance between the 10-cm mark and zero is off by 1 cm. The error in absolute position builds. Thus, the combination of these two effects leads to larger differences with respect to the experimental data for the thicker slabs.

5.4 Conclusions

As shown in the data above and the bond length results explained in [3], small changes in projected bond angles ($\approx 1^\circ$) can result in significant changes in atomic displacements without a simultaneous perceptible change in bond lengths. Therefore, one is unable simply to assume that the third TiO plane down from the surface has a bulk-like geometry and expect to obtain correct atomic displacements. This is due to the combined effects of vertical-relaxation errors for the deeper atoms and the relatively slow convergence of the bond angles responsible for vertical displacements of the surface atoms.

Surface atomic structures for this material and many other covalently bonded solids have typically been reported by specifying relaxed absolute atomic positions, which requires specifying an artificial reference plane. Since this has the potential to hide multiple different effects, as described here, we argue in favor of adopting a different approach for reporting surface structures, in which bond lengths and bond angles are specified instead of absolute atomic positions. This change would not only present the system in terms of more physically relevant quantities, but would also highlight the relative convergence rates of bond lengths and bond angles.

References

1. U. Diebold, Appl. Phys. A **76**, 681 (2003).
2. R. Lindsay, A. Wander, A. Ernst, B. Montanari, G. Thornton, N.M. Harrison, Phys. Rev. Lett. **94**, 246102 (2005).
3. S.J. Thompson, S.P. Lewis, Phys. Rev. B **73**, 073403 (2006).
4. G. Kresse, J. Furthmüller, Phys. Rev. B **54**, 11169 (1996).
5. G. Kresse, J. Furthmüller, Comput. Mater. Sci. **6**, 15 (1996).
6. R.M. Martin, *Electronic Structure Basic Theory and Practical Methods.* (Cambridge University Press, Cambridge 2004)
7. T. Bredow, L. Giordano, F. Cinquini, G. Pacchioni, Phys. Rev. B **70**, 035419 (2004).

Part II

New Models, Methods and Perspectives

6

Ensemble Optimization Techniques for the Simulation of Slowly Equilibrating Systems

S. Trebst[1], D.A. Huse[2], E. Gull[3], H.G. Katzgraber[3], U.H.E. Hansmann[4,5], and M. Troyer[3]

[1] Microsoft Research and Kavli Institute for Theoretical Physics, University of California, Santa Barbara, CA 93106, USA
[2] Department of Physics, Princeton University, Princeton, NJ 08544, USA
[3] Theoretische Physik, Eidgenössische Technische Hochschule Zürich, 8093 Zürich, Switzerland
[4] Department of Physics, Michigan Technological University, Houghton, MI 49931, USA
[5] John-von-Neumann Institute for Computing, Forschungszentrum Jülich, 52425 Jülich, Germany

Abstract. Competing phases or interactions in complex many-particle systems can result in free energy barriers that strongly suppress thermal equilibration. Here we discuss how extended ensemble Monte Carlo simulations can be used to study the equilibrium behavior of such systems. Special focus will be given to a recently developed adaptive Monte Carlo technique that is capable to explore and overcome the entropic barriers which cause the slow-down. We discuss this technique in the context of broad-histogram Monte Carlo algorithms as well as its application to replica-exchange methods such as parallel tempering. We briefly discuss a number of examples including low-temperature states of magnetic systems with competing interactions and dense liquids.

6.1 Introduction

The free energy landscapes of complex many-body systems with competing phases or interactions are often characterized by many local minima that are separated by entropic barriers. The simulation of such systems with conventional Monte Carlo [1] or molecular dynamics [2] methods is slowed down by long relaxation times due to the suppressed tunneling through these barriers. While at second order phase transitions this slow-down can be overcome by improved updating techniques, such as cluster updates [3, 4], this is not the case for systems which undergo a first-order phase transition or for systems that exhibit frustration or disorder. For these systems one instead aims at improving the way that relatively simple, local updates are accepted or rejected

Springer Proceedings in Physics, Volume 123
Computer Simulation Studies in Condensed-Matter Physics XIX
Eds.: D.P. Landau, S.P. Lewis and H.-B. Schüttler

in the sampling process by introducing artificial statistical ensembles such that tunneling times through barriers are reduced and autocorrelation effects minimized. In the following we discuss recently developed techniques to find statistical ensembles that optimize the performance of Monte Carlo sampling, first in the context of broad-histogram Monte Carlo algorithms and then outline how these methods can be applied in the context of replica-exchange or parallel-tempering algorithms.

6.2 Extended Ensemble Methods

Let us consider a first-order phase transition, such as in a two-dimensional q-state Potts model [5] with a Hamiltonian

$$H = -J \sum_{\langle i,j \rangle} \delta_{\sigma_i \sigma_j} , \tag{6.1}$$

where the spins σ_i can take the integer values $1, \ldots, q$. For $q > 4$ this model exhibits a first-order phase transition, accompanied by exponential slowing down of conventional local-update algorithms. The exponential slow-down is caused by the free-energy barrier between the two coexisting meta-stable states at the first-order phase transition.

This barrier can be quantified by considering the energy histogram

$$H_{\text{canonical}}(E) \propto g(E) P_{\text{Boltzmann}}(E) = g(E) \exp(-\beta E) , \tag{6.2}$$

which is the probability of encountering a configuration with energy E during the Monte Carlo simulation. The density of states is given by

$$g(E) = \sum_{c} \delta_{E,E(c)} , \tag{6.3}$$

where the sum runs over all configurations c. Away from first-order phase transitions, $H_{\text{canonical}}(E)$ has approximately Gaussian shape, centered around the mean energy. At first-order phase transitions, where the energy jumps discontinuously, the histogram $H_{\text{canonical}}(E)$ develops a double-peak structure. The minimum of $H_{\text{canonical}}(E)$ between these two peaks, which the simulation has to cross in order to go from one phase to the other, becomes exponentially small upon increasing the system size. This leads to exponentially large tunneling and autocorrelation times.

This tunneling problem at first-order phase transitions can be aleviated by extended ensemble techniques which aim at broadening the sampled energy space. Instead of weighting a configuration c with energy $E = E(c)$ using the Boltzmann weight $P_{\text{Boltzmann}}(E) = \exp(-\beta E)$ more general weights $P_{\text{extended}}(E)$ are introduced which define the extended ensemble. The configuration space is explored by generating a Markov chain of configurations

$$c_1 \to c_2 \to \ldots \to c_i \to c_{i+1} \to \cdots , \tag{6.4}$$

where a move from configuration c_1 to c_2 is accepted with probability

$$P_{\text{acc}}(c_1 \to c_2) = \min\left[1, \frac{P(c_2)}{P(c_1)}\right] = \min\left[1, \frac{W_{\text{extended}}(E_2)}{W_{\text{extended}}(E_1)}\right] . \tag{6.5}$$

In general, the extended weights are defined in a single coordinate, such as the energy, thereby projecting the random walk in configuration space to a random walk in energy space

$$E_1 = E(c_1) \to E_2 \to \ldots \to E_i \to E_{i+1} \to \cdots . \tag{6.6}$$

For this random walk in energy space a histogram can be recorded which has the characteristic form

$$H_{\text{extended}}(E) \propto g(E) W_{\text{extended}}(E) , \tag{6.7}$$

where the density of states $g(E)$ is fixed for the simulated system.

One choice of generalized weights is the multicanonical ensemble [6] where the weight of a configuration c is defined as $W_{\text{multicanonical}}(c) \propto 1/g(E(c))$. The multicanonical ensemble then leads to a flat histogram in energy space

$$H_{\text{multicanonical}}(E) \propto g(E) W_{\text{multicanonical}}(E) = g(E) \frac{1}{g(E)} = \text{const.} \tag{6.8}$$

removing the exponentially small minimum in the canonical distribution. After performing a simulation, measurements in the multicanonical ensemble are reweighted by a factor $W_{\text{Boltzmann}}(E)/W_{\text{multicanonical}}(E)$ to obtain averages in the canonical ensemble.

Since the density of states and thus the multicanonical weights are not known initially, a scalable algorithm to estimate these quantities is needed. The Wang-Landau algorithm [7] is a simple but efficient iterative method to obtain good approximates of the density of states $g(E)$ and the multicanonical weights $W_{\text{multicanonical}}(E) \propto 1/g(E)$. Besides overcoming the exponentially suppressed tunneling problem at first-order phase transitions, the Wang-Landau algorithm calculates the generalized density of states $g(E)$ in an iterative procedure. The knowledge of the density of states $g(E)$ then allows the direct calculation of the free energy from the partition function, $Z = \sum_E g(E) e^{-\beta E}$. The internal energy, entropy, specific heat and other thermal properties are easily obtained as well, by differentiating the free energy. By additionally measuring the averages $A(E)$ of other observables A as a function of the energy E, thermal expectation values can be obtained at arbitrary inverse temperatures β by performing just a single simulation:

$$\langle A(\beta) \rangle = \frac{\sum_E A(E) g(E) e^{-\beta E}}{\sum_E g(E) e^{-\beta E}} . \tag{6.9}$$

6.3 Markov Chains and Random Walks in Energy Space

The multicanonical ensemble and Wang–Landau algorithm both project a random walk in high-dimensional configuration space to a one-dimensional random walk in energy space where all energy levels are sampled equally often. It is important to note that the random walk in configuration space, (6.4), is a biased Markovian random walk, while the projected random walk in energy space, (6.6), is non-Markovian, as memory is stored in the configuration. This becomes evident as the system approaches a phase transition in the random walk: While the energy no longer reflects from which side the phase transition is approached, the current configuration may still reflect the actual phase the system has visited most recently. In the case of the ferromagnetic Ising model, the order parameter for a given configuration at the critical energy $E_c \sim -1.41N$ (in two space dimensions) will reveal whether the system is approaching the transition from the magnetically ordered (lower energies) or disordered side (higher energies).

This loss of information in the projection of the random walk in configuration space has important consequences for the random walk in energy space. Most strikingly, the local diffusivity of a random walker in energy space, which for a diffusion time t_D is defined as

$$D(E, t_D) = \langle (E(t) - E(t + t_D))^2 \rangle / t_D \tag{6.10}$$

is *not* independent of the location in energy space. This is illustrated in Fig. 6.1 for the two-dimensional Ising ferromagnet. Below the phase transition around $E_c \sim -1.41N$ a clear minimum evolves in the local diffusivity. In this region large ordered domains are formed and by moving the domain boundaries through local spin flips only small energy changes are induced resulting in a suppressed local diffusivity in energy space.

Because of the strong energy dependence of the local diffusivity the simulation of a multicanonical ensemble sampling all energy levels equally often turns out to be suboptimal [8]. The performance of flat-histogram algorithms can be quantified for classical spin models such as the ferromagnet where the number of energy levels is given by $[-2N, +2N]$ and thereby scales with the number of spins N in the system. When measuring the typical round-trip time between the two extremal energies for multicanonical simulations, these round-trip times τ are found to scale like

$$\tau \sim N^2 L^z , \tag{6.11}$$

showing a power-law deviation from the N^2-scaling behavior of a completely unbiased random walk. Here L is the linear system size and z a critical exponent describing the slow-down of a multicanonical simulation in the proximity of a phase transition [8, 9]. The value of z strongly depends on the simulated model and the dimensionality of the problem. In two dimensions the exponent increases from $z = 0.74$ for the ferromagnet as one introduces competing interactions leading to frustration and disorder. The exponent becomes $z = 1.73$

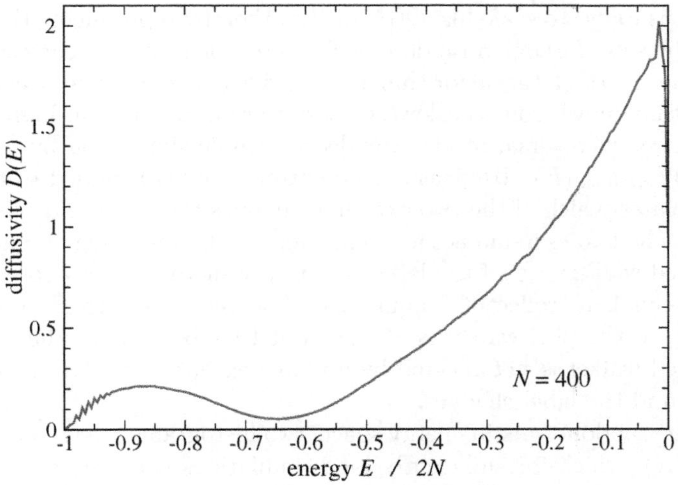

Fig. 6.1. Local diffusivity $D(E)$ of a random walk sampling a flat histogram in energy space for the two-dimensional ferromagnetic Ising model with $N = 20 \times 20$ spins. The local diffusivity strongly depends on the energy with a strong suppression around the critical energy $E_c \approx -1.41N$ and the ground-state energy $E_0 = -2N$

for the two-dimensional fully frustrated Ising model which is defined by a Hamiltonian

$$H = \sum_{\langle i,j \rangle} J_{ij}\sigma_i\sigma_j \,, \tag{6.12}$$

where the spins around any given plaquette of four spins are frustrated, e.g. by choosing the couplings along three bonds to be $J_{ij} = -1$ (ferromagnetic) and $J_{ij} = +1$ (antiferromagnetic) for the remaining bond. For the spin glass ,where the couplings J_{ij} are randomly chosen to be $+1$ or -1, exponential scaling $(z = \infty)$ is found [8, 10]. Increasing the spatial dimension of the ferromagnet the exponent is found to decrease as $z \approx 1.81, 0.74$ and 0.44 for dimension $d = 1, 2$ and 3 and z vanishes for the mean-field model in the limit of infinite dimensions [9].

6.4 Optimized Ensembles

The observed polynomial slow-down of the multicanonical ensemble poses the question whether for a given model there is an optimal choice of the histogram $H_{\text{optimal}}(E)$ and corresponding weights $W_{\text{optimal}}(E)$, which eliminates the slow-down. To address this question an adaptive feedback algorithm has recently been introduced that iteratively improves the weights in an extended ensemble simulation leading to further improvements in the efficiency of the algorithm by several orders of magnitude [11]. The scaling for the optimized ensemble is found to scale like $O([N \ln N]^2)$ thereby reproducing the behavior of an unbiased Markovian random walk up to a logarithmic correction.

At the heart of the algorithm lies the idea to maximize a current j of walkers that move from the lowest energy level, E_-, to the highest energy level, E_+, or vice versa, in an extended ensemble simulation by varying the weights $W_{\text{extended}}(E)$. To measure the current a label is added to the walker that indicates which of the two extremal energies the walker has visited most recently. The two extrema act as "reflecting" and "absorbing" boundaries for the labeled walker: e.g., if the label is plus, a visit to E_+ does not change the label, so this is a "reflecting" boundary. However, a visit to E_- does change the label, so the plus walker is absorbed at that boundary. The behavior of the labeled walker is *not* affected by its label except when it visits one of the extrema and the label changes.

For the random walk in energy space, two histograms are recorded, $H_+(E)$ and $H_-(E)$, which for sufficiently long simulations converge to steady-state distributions which satisfy $H_+(E) + H_-(E) = H(E) = W(E)g(E)$. For each energy level the fraction of random walkers which have label "plus" is then given by $f(E) = H_+(E)/H(E)$. The above-discussed boundary conditions dictate $f(E_-) = 0$ and $f(E_+) = 1$.

The steady-state current to first-order in the derivative is

$$j = D(E)H(E)\frac{df}{dE}, \tag{6.13}$$

where $D(E)$ is the walker's diffusivity at energy E. There is no current if $f(E)$ is constant, since this corresponds to the equilibrium state. Therefore the current is to leading order proportional to df/dE. Rearranging the above equation and integrating on both sides, noting that j is a constant and f runs from 0 to 1, one obtains

$$\frac{1}{j} = \int_{E_-}^{E_+} \frac{dE}{D(E)H(E)}. \tag{6.14}$$

To maximize the current and thus the round-trip rate, this integral must be minimized. However, there is a constraint: $H(E)$ is a probability distribution and must remain normalized which can be enforced with a Lagrange multiplier:

$$\int_{E_-}^{E_+} dE \left(\frac{1}{D(E)H(E)} + \lambda H(E) \right). \tag{6.15}$$

To minimize this integrand, the ensemble, that is the weights $W(E)$ and thus the histogram $H(E)$ are varied. At this point it is assumed that the dependence of $D(E)$ on the weights can be neglected.

The optimal histogram, $H_{\text{optimal}}(E)$, which minimizes the above integrand and thereby maximizes the current j is then found to be

$$H_{\text{optimal}}(E) \propto \frac{1}{\sqrt{D(E)}}. \tag{6.16}$$

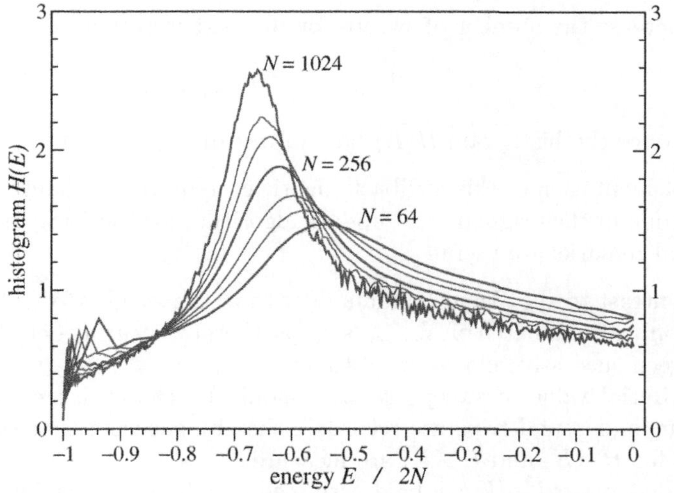

Fig. 6.2. Optimized histograms for the two-dimensional ferromagnetic Ising model for various system sizes. After the feedback of the local diffusivity a peak evolves near the critical energy of the transition $E_c \approx -1.41N$. The feedback thereby shifts additional resources towards the bottleneck of the simulation which are identified by a suppressed local diffusivity

Thus for the optimal ensemble, the probability distribution of sampled energy levels is simply inversely proportional to the square root of the local diffusivity.

The optimal histogram can be approximated in a feedback loop of the form

- Start with some trial weights $W(E)$, e.g. $W(E) = 1/g(E)$.
- Repeat
  - Reset the histograms $H(E) = H_+(E) = H_-(E) = 0$.
  - Simulate the system with the current weights for N sweeps:
    - Updates are accepted with probablity $\min[1, W(E')/W(E)]$.
    - Record the histograms $H_+(E)$ and $H_-(E)$.
  - From the recorded histogram an estimate of the local diffusivity is obtained as

$$D(E) \propto \frac{1}{H(E)\frac{df}{dE}} ,$$

where the fraction $f(E)$ is given by $f(E) = H_+(E)/H(E)$ and $H(E)$ is the histogram $H(E) = H_+(E) + H_-(E)$.

  - Define new weights as

$$W_{\text{optimized}}(E) = W(E)\sqrt{\frac{1}{H(E)} \cdot \frac{df}{dE}} .$$

– Increase the number of sweeps for the next iteration

$$N_{\text{sweeps}} \leftarrow 2N_{\text{sweeps}} .$$

- Stop once the histogram $H(E)$ has converged.

The implementation of this feedback algorithm requires to change only a few lines of code in the original local-update algorithm for the Ising model. Some additional remarks are useful:

1. In contrast to the Wang-Landau algorithm, the weights $W(E)$ are modified only after a batch of N_{sweeps} sweeps, thereby ensuring detailed balance between successive moves at all times.
2. The initial value of sweeps N_{sweeps} should be chosen large enough that a couple of round trips are recorded, thereby ensuring that steady state data for $H_+(E)$ and $H_-(E)$ are measured.
3. The derivative df/dE can be determined by a linear regression, where the number of regression points is flexible. Initial batches with the limited statistics of only a few round trips may require a larger number of regression points than subsequent batches with smaller round-trip times and better statistics.
4. Similar to the multicanonical ensemble, the weights $W(E)$ can become very large and storing the logaritms may be advantageous. The reweighting then becomes $\ln W_{\text{optimized}}(E) = \ln W(E) + [\ln \frac{df}{dE} - \ln H(E)]/2$.

At the end of the simulation, the density of states can be estimated from the recorded histogram as $g(E) = H_{\text{optimized}}(E)/W_{\text{optimized}}(E)$ and normalized as described above.

Figure 6.2 shows the optimized histogram for the two-dimensional ferromagnetic Ising model. The optimized histogram is no longer flat, but a peak evolves at the critical region around $E_c \approx -1.41N$ of the transition. The feedback of the local diffusivity reallocates resources towards the bottlenecks of the simulation which have been identified by a suppressed local diffusivity.

When analyzing the scaling of round-trip times for the optimized ensemble one finds a considerable speedup: The power-law slow-down of round-trip times for the flat-histogram ensemble $O(N^2L^z)$ is reduced to $O([N \ln N]^2)$ for the optimized ensemble, e.g. there is only a logarithmic correction to the scaling of a completely unbiased random walk with $O(N^2)$-scaling. For the two-dimensional fully frustrated Ising model the scaling of round-trip times is shown in Fig. 6.3. This scaling improvement results in a speedup by a nearly two orders of magnitude already for a system with some 128×128 spins.

6.5 Simulation of Dense Fluids

Extended ensembles cannot only be defined as a function of energy, but in arbitrary reaction coordinates $\boldsymbol{R}$ onto which a random walk in configuration

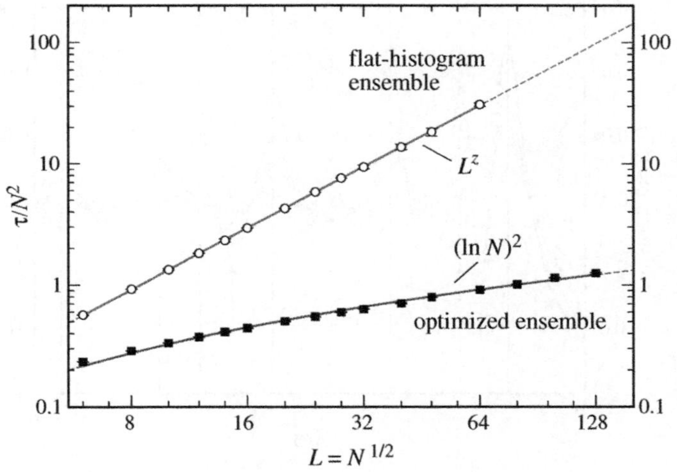

Fig. 6.3. Scaling of round-trip times for a random walk in energy space sampling a flat histogram (*open squares*) and the optimized histogram (*solid circles*) for the two-dimensional fully frustrated Ising model. While for the multicanonical simulation a power-law slow-down of the round-trip times $O(N^2 L^z)$ is observed, the round-trip times for the optimized ensemble scale like $O([N \ln N]^2)$ thereby approaching the ideal $O(N^2)$-scaling of an unbiased Markovian random walk up to a logarithmic correction

space can be projected. The generalized weights in these reaction coordinates $W_{\text{extended}}(\boldsymbol{R})$ are then used to bias the random walk along the reaction coordinate by accepting moves from a configuration c_1 with reaction coordinate $\boldsymbol{R}_1$ to a configuration c_2 with reaction coordinate $\boldsymbol{R}_2$ with probability

$$p_{\text{acc}}(c_1 \to c_2) = p_{\text{acc}}(\boldsymbol{R}_1 \to \boldsymbol{R}_2) = \min\left(1, \frac{W_{\text{extended}}(\boldsymbol{R}_2)}{W_{\text{extended}}(\boldsymbol{R}_1)}\right) . \tag{6.17}$$

The generalized weights $W_{\text{extended}}(\boldsymbol{R})$ can again be chosen in such a way that similar to a multicanonical simulation a flat histogram is sampled along the reaction coordinate by setting the weights to be inversely proportional to the density of states defined in the reaction coordinates, that is $W_{\text{extended}}(\boldsymbol{R}) \propto 1/g(\boldsymbol{R})$.

An optimal choice of weights can be found by measuring the local diffusivity of a random walk along the reaction coordinates and by applying the feedback method to shift weight towards the bottlenecks in the simulation. This generalized ensemble optimization approach has recently been illustrated for the simulation of dense Lennard-Jones fluids close to the vapor-liquid equilibrium [12]. The interaction between particles in the fluid is described by a pairwise Lennard-Jones potential of the form

$$\Phi_{\text{LJ}}(R) = 4\epsilon\left[\left(\frac{\sigma}{R}\right)^{12} - \left(\frac{\sigma}{R}\right)^{6}\right] , \tag{6.18}$$

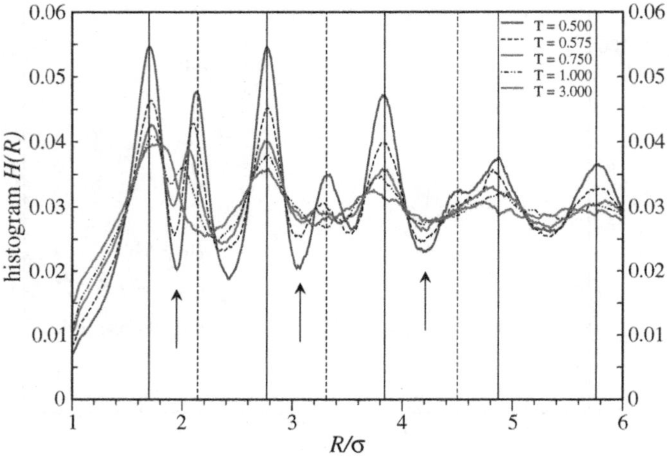

Fig. 6.4. Optimized histograms for a dense two-dimensional Lennard–Jones fluid after feedback of the local diffusivity in a radial coordinate for varying temperatures. For the optimized ensemble peaks evolve at the free energy barriers between the shells of the liquid which proliferate at lower temperatures (*solid lines*). Additional peaks emerge at low temperatures (*dashed lines*) revealing interstitial states (*arrows*) between the shells of the liquid

where ϵ is the interaction strength, σ a length parameter, and R the distance between two particles. It is this distance R between two arbitrarily chosen particles in the fluid that one can use as a new reaction coordinate for a projected random walk. For a given temperature defining an extended ensemble with weights $W_{\text{extended}}(R)$ and recording a histogram $H(R)$ during a simulation will then allow to calculate the pair distribution function $g(R) = H(R)/W_{\text{extended}}(R)$. The pair distribution function $g(R)$ is closely related to the potential of mean force (PMF)

$$\Phi_{\text{PMF}}(R) = -\frac{1}{\beta} \ln g(R) , \tag{6.19}$$

which describes the average interaction between two particles in the fluid in the presence of many surrounding particles.

For high particle densities and low enough temperatures shell structures form in the fluid which are reminiscent of the hexagonal lattice of the solid structure at very low temperatures. These shell structures are revealed by a sinusoidal modulation in the PMF. Thermal equilibration between the shells is suppressed by entropic barriers which form between the shells. Again, one can ask what probability distribution, or histogram, should be sampled along the reaction coordinate, in this case the radial distance R, so that equilibration between the shells is improved. Measuring the local diffusivity for a random walk along the radial distance R in an interval $[R_{\text{min}}, R_{\text{max}}]$ and subsequently applying the feedback algorithm described above optimized histograms $H(R)$

are found which are plotted in Fig. 6.4 for varying temperatures [12]. The feedback algorithm again shifts additional weight in the histogram towards the bottleneck of the simulation, in this case towards the barriers between the shells. Interestingly, additional peaks emerge in the optimized histogram as the temperature is lowered towards the vapor-liquid equilibrium. The minima between these peaks point to additional meta-stable configurations which occur at these low temperatures, namely interstitial states which occur as the shells around two particles merge as detailed in [12].

This example illustrates that for some simulations the local diffusivity and optimized histogram *themselves* are very sensitive measures that can reveal interesting underlying physical phenomena which are otherwise hard to detect in a numerical simulation. In general, a strong modulation of the local diffusivity for the random walk along a given reaction coordinate is a good indicator that the reaction coordinate itself is a good choice that captures some interesting physics of the problem.

6.6 Parallel Tempering / Replica-Exchange Methods

The simulation of frustrated and/or disordered systems suffers from a similar tunneling problem than the simulation of first-order phase transitions: local minima in energy space are separated by barriers that grow with system size. While the multicanonical or optimized ensembles do not help with the NP-hard problems faced by spin glasses, they are efficient in speeding up simulations of frustrated magnets without disorder [11].

An alternative to these extended ensembles for the simulation of frustrated magnets is the "parallel tempering" or "replica-exchange" Monte Carlo method [13–16]. Instead of performing a single simulation at a fixed temperature, simulations are performed for M replicas at a set of temperatures $T_1, T_2, \ldots, T_M$. In addition to standard Monte Carlo updates at a fixed temperature, exchange moves are proposed to swap two replicas between adjacent temperatures. These swaps are accepted with a probability

$$\min[1, \exp(\Delta\beta \Delta E)] , \tag{6.20}$$

where $\Delta\beta = \beta_j - \beta_i$ is the difference in inverse temperatures and $\Delta E = E_j - E_i$ the difference in energy between the two replicas i and j.

The effect of these exchange moves is that a replica can drift from a local free energy minimum at low temperatures to higher temperatures, where it is easier to cross energy barriers and equilibration is fast. Upon cooling (by another sequence of exchanges) it can end up in a different local minimum on time scales that are much shorter compared to a single simulation at a fixed low temperature. This random walk of a single replica in temperature space is the conjugate analog of the random walk in energy space discussed for the extended ensemble techniques. The complement of the statistical ensemble, defined by the weights $W_{\mathrm{extended}}(E)$, is the particular choice

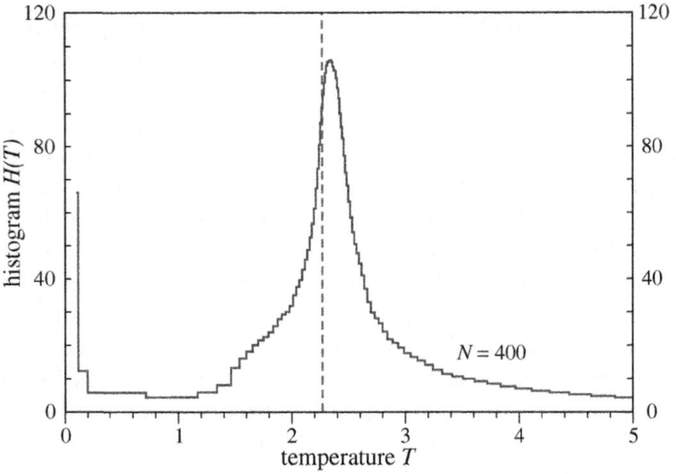

Fig. 6.5. Optimized temperature distribution $H(T)$ for a two-dimensional Ising ferromagnet with $N = 20 \times 20$ spins and 100 replicas/temperature points. After the feedback of the local diffusivity a peak evolves near the critical temperature of the transition $T_c \approx 2.269$. The feedback thereby shifts additional resources towards the bottleneck of the simulation which are identified by a suppressed local diffusivity. Note the similarity to Fig. 6.2

of temperature points in the temperature set $\{T_1, T_2, \ldots, T_M\}$ for the parallel tempering simulation. The probability of sampling any given temperature T in an interval $T_i < T < T_{i+1}$ can then be approximated by $H(T) \propto 1/\Delta T$, where $\Delta T = T_{i+1} - T_i$ is the length of the temperature interval around the temperature T. This probability distribution $H(T)$ is the equivalent to the histogram $H(E)$ in the extended ensemble simulations. The ensemble optimization technique discussed above can thus be reformulated to optimize the temperature set in a parallel-tempering simulation in such a way that the rate of round trips between the two extremal temperatures, T_1 and T_M respectively, is maized [17, 18].

Starting with an initial temperature set $\{T_1, T_2, \ldots, T_M\}$ a parallel tempering simulation is performed where each replica is labeled either "plus" or "minus" indicating which of the two extremal temperatures the respective replica has visited most recently. This allows to measure a current of replicas diffusing from the highest to the lowest temperature by recording two histograms, $h_+(T)$ and $h_-(T)$ for each temperature point. The current j is then given by

$$j = D(T)H(T)\frac{df}{dT} , \tag{6.21}$$

where $D(T)$ is the local diffusivity for the random walk in temperature space, and $f(T) = h_+(T)/[h_+(T) + h_-(T)]$ is the fraction of random walkers which have visited the highest temperature T_M most recently. The probability dis-

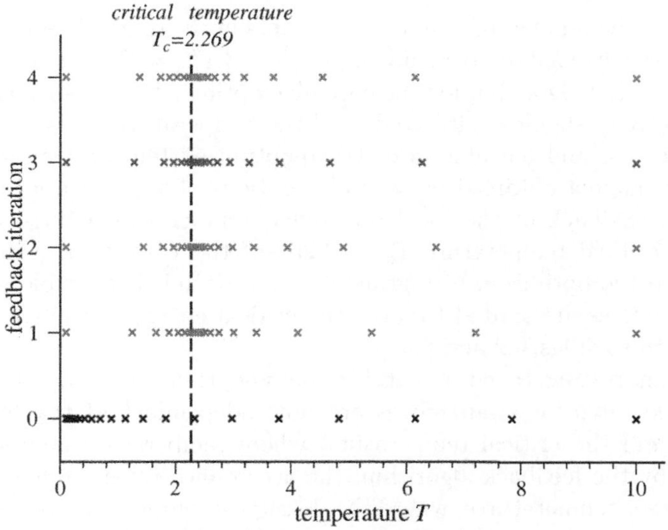

Fig. 6.6. Optimized temperature sets for a two-dimensional Ising ferromagnet with $N = 20 \times 20$ spins. The initial temperature set with 20 temperature points is determined by a geometric progression [17, 19] for the temperature interval [0.1, 10]. After feedback of the local diffusivity the temperature points accumulate near the critical temperature $T_c = 2.269$ of the phase transition (*dashed line*). Similar to the ensemble optimization in energy space the feedback of the local diffusivity relocates resources towards the bottleneck of the simulation

tribution $H(T)$ is normalized, that is

$$\int_{T_1}^{T_M} H(T)\, dT = C \int_{T_1}^{T_M} \frac{dT}{\Delta T} = 1 \,, \tag{6.22}$$

where C is a normalization constant. Rearranging (6.21), the local diffusivity $D(T)$ of the random walk in temperature space can be estimated as

$$D(T) \propto \frac{\Delta T}{df/dT} \,. \tag{6.23}$$

In analogy to the argument for the extended ensemble in energy space the current j is maximized by choosing a probability distribution

$$H_{\text{optimal}}(T) \propto \frac{1}{\sqrt{D(T)}} \propto \sqrt{\frac{1}{\Delta T}\frac{df}{dT}} \,, \tag{6.24}$$

which is inversely proportional to the square root of the local diffusivity. The optimized temperature set $\{T'_1, T'_2, \ldots, T'_M\}$ is then found by choosing the n-th temperature point T'_n such that

$$\int_{T'_1}^{T'_n} H_{\text{optimal}}(T)\, dT = \frac{n}{M} \,, \tag{6.25}$$

where M is the number of temperature points in the original temperature set, and the two extremal temperatures $T'_1 = T_1$ and $T'_M = T_M$ remain unchanged. Similarly to the algorithm for the ensemble optimization this feedback of the local diffusivity should be iterated until the temperature set is converged.

Figures 6.5 and 6.6 illustrate the so-optimized temperature sets for the Ising ferromagnet obtained by several iterations of the above feedback loop. After the feedback of the local diffusivity, temperature points accumulate near the critical temperature $T_c = 2.269$ of the transition. This is in full analogy to the optimized histograms for the extended ensemble simulations where resources are shifted towards the critical energy of the transition, for comparison see Figs. 6.2 and 6.5.

It is interesting to note that for the so-optimized temperature set the acceptance rates for swap moves are not independent of the temperature [17]. Around the critical temperature, where temperature points are accumulated by the feedback algorithm, the acceptance rates are higher than at higher/lower temperatures, where the density of temperature points becomes considerably smaller after feedback [17,18]. The almost Markovian scaling behavior for the optimized random walks in either energy or temperature space is thus generated by a problem-specific statistical ensemble which is characterized neither by a flat histogram nor flat acceptance rates for exchange moves, but by a characteristic probability distribution which concentrates resources at the minima of the measured local diffusivity.

6.7 Outlook

The ensemble optimization technique reviewed in this chapter should be broadly applicable to a wide range of applications – possibly speeding up existing uniform sampling techniques by orders of magnitude. It has recently been used to improve broad-histogram Monte Carlo techniques [11] as well as parallel-tempering Monte Carlo simulations [17], with applications to frustrated and disordered spin systems [11,17], dense fluids [12], as well as folded proteins [18]. It also holds promise to improve the simulation of quantum systems close to a phase transition when optimizing the extended ensemble introduced for the quantum Wang–Landau algorithm outlined in [20].

References

1. D.P. Landau, K. Binder, *A guide to Monte Carlo Simulations in Statistical Physics.* (Cambridge University Press, 2000)
2. D. Frenkel, B. Smit, *Understanding Molecular Simulation.* (Academic Press, 1996)
3. R.H. Swendsen, J.-S. Wang, Phys. Rev. Lett. **58**, 86 (1987).
4. U. Wolff, Phys. Rev. Lett. **62**, 361 (1989).

5. F.Y. Wu, Rev. Mod. Phys. **54**, 235 (1982).
6. B.A. Berg, T. Neuhaus, Phys. Lett. B **267**, 249 (1991); Phys. Rev. Lett. **68**, 9 (1992).
7. F. Wang, D.P. Landau, Phys. Rev. Lett. **86**, 2050 (2001); Phys. Rev. E **64**, 056101 (2001).
8. P. Dayal, S. Trebst, S. Wessel, D. Würtz, M. Troyer, S. Sabhapandit, S.N. Coppersmith, Phys. Rev. Lett. **92**, 097201 (2004).
9. Y. Wu, M. Körner, L. Colonna-Romano, S. Trebst, H. Gould, J. Machta, M. Troyer, Phys. Rev. E **72**, 046704 (2005).
10. S. Alder, S. Trebst, A.K. Hartmann, M. Troyer, J. Stat. Mech. P07008 (2004).
11. S. Trebst, D.A. Huse, M. Troyer, Phys. Rev. E **70**, 046701 (2004).
12. S. Trebst, E. Gull, M. Troyer, J. Chem. Phys. **123**, 204501 (2005).
13. R.H. Swendsen, J. Wang, Phys. Rev. Lett. **57**, 2607 (1986).
14. E. Marinari, G. Parisi: Europhys. Lett. **19**, 451 (1992).
15. A.P. Lyubartsev, A.A. Martsinovski, S.V. Shevkunov, P.N. Vorontsov-Velyaminov, J. Chem. Phys. **96**, 1776 (1992).
16. K. Hukushima, Y. Nemoto, J. Phys. Soc. Jpn. **65**, 1604 (1996).
17. H.G. Katzgraber, S. Trebst, D.A. Huse, M. Troyer, J. Stat. Mech. P03018 (2006).
18. S. Trebst, M. Troyer, U.H.E. Hansmann, J. Chem. Phys. **124**, 174903 (2006).
19. C. Predescu, M. Predescu, C.V. Ciobanu, J. Chem. Phys. **120**, 4119 (2004).
20. M. Troyer, S. Wessel, F. Alet, Phys. Rev. Lett. **90**, 120201 (2003).

7

The Avogadro Challenge

N. Ito

Department of Applied Physics, Graduate School of Engineering,
The University of Tokyo, Hongo 7-3-1, Bunkyo-ku, Tokyo 113-8656, Japan

Abstract. The Avogadro challenge is to reconstruct and analyze macroscopic behavior based on microscopic dynamics with the help of computer. It now reaches to the stage to reproduce transport phenomena like heat and electric conduction, and Newtonian fluid. It's applications have started from nanoscopic to mesoscopic sciences and technologies.

7.1 One mol Calculations

Remarkable developments in computing physics have been driven by exponential performance growth of computers. The famous Moore's law [1] implies that speed of one processor becomes two times faster at every eighteen months. Not only processors, but also peripheral technologies grow faster. Disk storage capacity becomes two times larger at every twelve months and network bandwidth is doubled at every nine months [2]. Statistical physics has been catching up the computer developments. This will be confirmed by the fact that speed and a scale of the Ising simulation with naive single-spin flip dynamics grows ten times at every four years [3, 4]. And computer simulation became an indispensable tool of statistical physics since 1980s as it is observed in the estimation history of the dynamical exponent [5] (see also Fig. 4.14 in [6]), for example. The dynamical exponent of square Ising model has been studied for long, but estimated values of it had been scattered in between 1.8 and 2.2, or more widely. However, estimations from computer simulations had been steadily converging and the ambiguity became less than 1% [5].

Present arithmatic speed of one system is going over 300 TFLOPS [7], and we are expecting 1 PFLOPS and more. One ambitious project is the KEISOKU computer project by RIKEN, which is planning to develop 10 PFLOPS system in five year with $1.25 \times 10^1 1$ yen. "KEISOKU" means "$10^1 6$ speed" in Japanese. This sounds ambitious, but it is safely in scope of the present computer technologies. TFLOPS graphics-oriented processor is

Springer Proceedings in Physics, Volume 123
Computer Simulation Studies in Condensed-Matter Physics XIX
Eds.: D.P. Landau, S.P. Lewis and H.-B. Schüttler

now available at hundreds dollars, that is, Microsoft's Xbox 360 or SONY's PS3. Therefore price of "KEISOKU game machine" is millions of dollars. With hundred times more budget, more general purpose computer will be built.

What is the most characteristic feature of the 10 PFLOPS computer? For most computing fields, 10 P computer may be just hundreds times faster computer. So it will improve integration time interval, mesh resolution, or accuracy. In statistical physics, however, 10 PFLOPS has a special importance. This is because the 10 P computer is the first machine which realize the Avogadro scale operations. During its machine life-time, which will be a few years, total number of floating operations (FLOP) may reach 10^{16} FLOPS $\times 10^8$ s ($\approx$ 3 years) $= 10^{24} = 1.7$ mol, and exceed the Avogadro number. Avogadro number is a number characterize macroscopic system based on microscopic elements. In this sense, the 10 PFLOPS computer will be a symbolic of an epoch in statistical physics. The Avogadro number has been a synonym of infinity. It is actually just one number, but it has been too big to enumerate. Now it is becoming just one number, and we can treat Avogadro scale procedure. Computer is having power to analyze macroscopic phenomena using microscopic dynamics, which has been the ultimate goal of statistical physics.

With above perspective, one of the most fascinating challenges will be to reconstruct macroscopic phenomena using only microscopic models in computer and analyze it. Let's call it the "Avogadro challenge". The purpose of this article is to review recent developments in the Avogadro challenge, mainly concentrated on the basic understanding of macroscopic behavior, not on application oriented way. Recently, so-called the "multiscale simulation" and "multiscale physics" have become popular, and the output from the Avogadro challenge will help such studies.

The first step of Avogadro challenge is to analyze the thermal equilibrium state. It has been one of the main problems of computing statistical physics, and big success has been achieved [6, 8, 9], and its applications to theoretical material design are now a very active field.

The Ising model has been a standard model in statistical physics [10–12]. Many simulation studies has been devoted for it and now the simulaion scale has reached almost macroscopic scale. Recent nonequilibrium relaxation study of cubic Ising model used up to $6648 \times 6648 \times 6656$ lattice to evaluate the critical temperature with 2.3 ppm relative error [13][1]. Such size corresponds to μm^3 or larger crystal which is bigger than domain recording one bit in present magnetic hard disk device(see also Fig. 7.1), so it is already a macroscopic system even in experimental sense. In this scale, nonequilibrium behavior with dislocation, domain, wall, interface and so on is also crucial to the description of the system.

In contrast to the success in the equilibrium state, nonequilibrium statistical physics still remains in unsatisfactory level. When one leaves the equilib-

[1] If only memory capacity is considered, $88,000^3$ Ising spin is simulated with 10 TB memory which is available on present big computer.

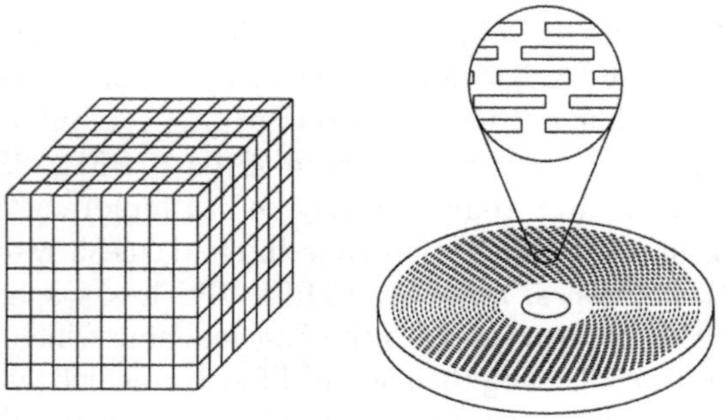

Fig. 7.1. 6000^3 Ising model (*left*) is larger than one-bit recording domain in hard-disk device (*right*). Now nonequilibrium simulation becomes more crucial task

rium state, big obstacles appear immediately in linear nonequilibrium regime. Linear response theory and the Kubo formula give expression of transport coefficient c with autocorrelation function $C(t)$ in equilibrium [14] as

$$c \propto \int_0^\infty C(t)\mathrm{d}t\,, \tag{7.1}$$

but estimation of the autocorrelation functions are usually difficult. Furthermore, they often show power-law decay known as the long-time tails [15] and integration in the Kubo formula sometimes seem to diverge. Typical long-time tail behavior is

$$C(t) \propto t^{-d/2}\,, \tag{7.2}$$

where d denotes the system dimensionality. This behavior is interpreted as a consequence of diffusion process or of hydrodynamic interaction. With this power-law decay (7.2), transport coefficient from Kubo formula (7.1) shows power-law divergence in one-dimensional system, logarithmic divergence in two-dimensional system, and power-law convergence in three-dimensional system:

$$c \sim \begin{cases} T^{1/2} \\ \log T \\ c_0 + c_1 T^{-1/2} \end{cases}, \quad T \to \infty\,. \tag{7.3}$$

It is remarked that slow power-law convergence is expected even in three-dimensional system. Assuming that the cut-off time denoted by T is propotional to the system size L, this anomaly is expected to be observed as size dependence of the transport coefficients of finite systems as [16]

$$c \sim \begin{cases} L^{1/2} \\ \log L \\ c_0 + c_1 L^{-1/2} \end{cases}. \tag{7.4}$$

Computer simulation studies had also been failed to reproduce proper transport. Since Fermi, Pasta and Ulam challenged simulation of heat conduction phenomena [17], many tried with various models, and only some tricky nonlinear chain systems which locally violate momentum conservation law had been shown finite transport coefficients [18]. Size dependence of momentum conserving models were confirmed to show power-law divergence expected from (7.4), although the exponent is smaller than 0.5 [18]. And logarithmic divergence was confirmed in two-dimensional lattice system [19]. Threrefore the size depencence (7.4) seems to be reliable and finite transport coefficient has been expected in three-dimensional system, although it had not been confirmed.

Recent studies, however, succeeded to reproduce proper transport phenomena in linear nonequilibrium regime. Since finite heat conductivity were observed in three-dimensional momentum conserving models [16, 22], other transport phenomena have been reproduced in computer simualtions. In the following sections, simulations of heat conductivity [16, 20–22], electric conductivity [23] and shear viscosity [24, 25] are reviewed. Statistical physical simulations of nonequilibrium structures are also reviewed [20, 21, 26, 27].

7.2 Heat Conductivity

Finite heat conductivity was firstly confirmed in hard-sphere system and in a nonlinear lattice model [16, 21]. It was also confirmed in Lennard-Jones particle system [22, 29].

7.2.1 Hard-sphere System

Nonequilbrium particle dynamics simulation of hard-spheres in three-dimensional parallelpiped box with high and low temperature walls in left and right ends, respectively, reproduced Fourier-type linear temperature profile after initial relaxation process and ratio of energy flux to temperature gradient converged when the system size is increased. The convergence speed was consistent with $L^{-0.5}$ from (7.4). Radius and mass of hard spheres were taken to be 0.1 and 1. Simulation box was $1 \times 1 \times L_z$, and L_z was changed from 2 to 20. Boundary condition to L_z directions were thermal walls, that is, a sphere forgot its velocity and is assigned new velocity randomly from Boltzmann distribution of the wall temperature at every collision with thermal walls. The perpendicular directions have periodic boundary conditions. System density was less than the freezing density all over the system. Long time tail behavior, $t^{-1.5}$, of (7.2) was also observed in heat-flux autocorrelation function.

Heat conductivity of system with $L_z = 7$ was already within 10% of the converging value. If we interpret the sphere as a molecule of size of 1 nm, $L_z = 7$ corresponds 70 nm. Smaller than this length scale, conductivity and dissipation accompanied with transport decrease as $1/\sqrt{L}$. Ballistic transport becomes dominant in smaller size.

Convergence was confirmed also in solid phase. Two-dimensional system was also studied and t^{-1} tail was observed.

7.2.2 Nonlinear lattice model

Another three-dimensional system confirmed to have finite heat conductivity was a nonlinear lattice. Interparticle potential had biquadratic term in addition to harmonic term, that is,

$$V_{ij} = \frac{1}{2}(\Delta q_{ij})^2 + \frac{g}{4}(\Delta q_{ij})^4 , \tag{7.5}$$

where

$$\Delta q_{ij} = |\boldsymbol{q}_i - \boldsymbol{q}_j| - l_0 , \tag{7.6}$$

$\boldsymbol{q}_i$ denotes coordinate vector of particle i and l_0 is a positive constant. This is so-called the FPU β model. It is remarked that the "natual length" l_0 should be added in three-dimensional lattice, although it is not relevant in one-dimensional lattice because a linear transformation of coordinates absorbs constant l_0.

Simple cubic lattice of the size of $3 \times 3 \times N_z$ was studied with N_z from 8 to 256. Nos'e-Hoover thermostats with high and low temperature were charged to particles in both ends to N_z direction, and boundary condition to N_z direction is free. Periodic boundaries are charged to perpendicular directions to N_z. Although this system is three-dimensional nonlinear lattice, it is not like crystal, but like polymer chain [28]. Finite heat conductivity in the limit of $N_z \to \infty$ was observed for this system.

In nonlinar lattice model case, above system may be exceptional, and most system tends to show logarithmic(or stronger, perhaps) divergence even in three or higher dimension [28] as far as the system size observed in present computer simulations. This may be because the contribution to energy flow from ballistic mode is not negligible. Separate treatment of heat flux from energy flux is necessary and naive analysis applied so far is not sufficient.

7.2.3 Lennard-Jones System

Particle system obeys (7.4), and shows finite conductivity in rather small system, in contrast to nonlinear lattices. Nonequilibrium simulation for system of particles with Lennard-Jones interaction,

$$V_{ij} = 4\epsilon \left\{ \left(\frac{\sigma}{q_{ij}} \right)^{12} - \left(\frac{\sigma}{q_{ij}} \right)^{6} \right\} , \tag{7.7}$$

confirmed $L^{-0.5}$ convergence of heat conductivity in all phases, that is, solid, liquid, gas and supercritical phases. Heat conductivity in supercritical phase turned out to be proportional to the density.

7.3 Other Transport Phenomena

7.3.1 Electric Conductivity

The results of heat conductivity can be applied to another important transport phenomena, electric conductivity. One model simple but realistic model of Ohmic conductor was proposed and studied [23]. This model consists three kinds of elastic particles: charge carrier, phonon and impurity. They all obey classical equation of motion and they all have the same radius and the same Young modulus. External electric field acts force only on charge carrier. Charge carrier does not feel heat bath, but phonon particle thermalizes at the boundary. Impurities are located at randomly with prescribed density, and they do not move.

The system is two dimensional, but impurity is considered to suppress the divergence of electric conductivity. Ohmic linear response in electric current to external field was confirmed, and fluctuation-dissipation relation and Kramers-Kronig relation were also confirmed.

7.3.2 Viscosity

Another important transport phenomena will be hydrodynamic viscosity. It is shown that hard-disk system reproduces two-dimensional Newtonian fluid [24]. The system is two dimensional, and therefore the viscosity is considered to diverge logarithmically in large limit. This divergence will be, however, useful in this system because it helps to adjust the Reynolds number. Poiseuille flow and flow passed cylinder is reproduced. Three-dimensional hard-disk system is also studied [25].

7.4 Nonequilibrium Structure

We have seen briefly how the statistical physical simulations reproduce linear nonequilibrium macroscopic behavior. If one designs models and simulations carefully, macroscopic behavior is reproduced with rather small number of particles. Heat conductions were already reproduced with thousands or tens thousands particles. Such small system in actual world corresponds to tens or hundreds nanometer system, and therefore results from such simulations will be useful in the nanoscale sciences and technologies. Furthermore, it will imply that the potential of the Avogadro Challenge.

Challenges beyond transport properties has started. Multiphase system with heat and hydrodynamics flows, phase transition was reproduced in one system simultaneously with hard particles [20, 21], or Lennard-Jones particles [27]. Structure of solid-fluid and liquid-gas interfaces are studied [21, 27]. Karman vortices and Taylor vortices were reproduced [21, 24] Nanobubble formation caused by heat pulse and evaporation was also studied [26].

Statistical physical analysis on nonequilibrium reaction system showed that the behavior of the reaction rate of reaction network changes from Arrhenius behavior near equilibrium state to power law regime when the driving force exceeds some threshold [30, 31].

7.5 Perspective

The idea and the present status of the Avogadro Challenge are given in this article. It is shown that the transport phenomena are simulated properly based mainly on microscopic dynamics and with an appropriate simulation box and boundary conditions. Such simulations will be useful in the studies of nonequilibrium science and technologies. Now the computer simulation has been becoming a tool not just to simulate the system, but to emulate with the actual phenomena. In this sense, the Avogadro Challenge will be the first step of the "computer emulation physics".

References

1. G.E. Moore, Electronics **38**, 8 (1965). This article and further details are available in http://www.intel.com/technology/mooreslaw/
2. I. Foster, Physics Today **55**, 35 (2002).
3. N. Ito, "Monte Carlo Study of the Ising Model" (Dissertation, The University of Tokyo, 1990).
4. N. Ito, *Computer Simulation Studies in Condensed-Matter Physics XI.* D.P. Landau, H.-B. Schüttler (Eds.) (Springer, Berlin Heidelberg New York 1999) p. 130.
5. N. Ito, Physica A **196**, 569 (1993).
6. D.P. Landau, K. Binder, *A Guide to Monte Carlo Simulations in Statistical Physics.* 2nd Edn. (Cambridge University Press, 2005).
7. http://www.top500.org/lists/2005/11/basic
8. K. Binder (Ed.), *Monte Carlo Methods in Statistical Physics.* (Springer, Berlin Heidelberg New York, 1979).
9. K. Binder (Ed.), *Applications of the Monte Carlo Methods in Statistical Physics.* 2nd Edn. (Springer, Berlin Heidelberg New York 1987).
10. W. Lentz, Z. Physik **21**, 613 (1920).
11. E. Ising, Z. Physik **31**, 253 (1925).
12. S.G. Brush, Rev. Mod. Phys. **39**, 883 (1967).
13. N. Ito, Pramana-Journal of Physics (Indian Academy of Sciences) vol. 64 No.5&6 (2005) p. 871–880 as Proc. of the 22nd Internaltional Conference on Statistical Physics (Bangalore, India 2005).
14. R. Kubo, J. Phys. Soc. Jpn. **12**, 570 (1957).
15. B.J. Alder, T.E. Wainwright, Phys. Rev. A **1**, 18 (1970).
16. T. Shimada, T. Murakami, S. Yukawa, K. Saito, N. Ito, J. Phys. Soc. Jpn. **69**, 3150 (2000).
17. E. Fermi, J. Pasta, S. Ulam, Los Alamos Report LA1940 (1955).

18. S. Lepri, R. Livi, A. Politi, Phys. Rev. **377**, 1 (2003).
19. A. Lippi, R. Livi, J. Stat. Phys. **100**, 1147 (2000).
20. T. Murakami, S. Yukawa, N. Ito, *Computer Simulation Studies in Condensed Matter Physics XIV.* D.P. Landau, S.P. Lewis, H.B. Schüttler (Eds.), (Springer, Berlin Heidelberg New York 2002).
21. T. Murakami, T. Shimada, S. Yukawa, N. Ito, J. Phys. Soc. Jpn. **72**, 1049 (2003).
22. F. Ogushi, S. Yukawa, N. Ito, J. Phys. Soc. Jpn. **74**, 827 (2005).
23. T. Yuge, A. Shimizu, N. Ito, J. Phys. Soc. Jpn. **74**, 1895 (2005).
24. T. Ishiwata, T. Murakami, S. Yukawa, N. Ito, Intern. J. Modern. Phys. C **15**, 1413 (2004).
25. T. Murakami, *Simulational Study of Nonequilibrium Phenomena.* (Dissertation, The University of Tokyo, 2003).
26. H. Okumura, N. Ito, Phys. Rev. E **67**, 045301 (2003).
27. F. Ogushi, S. Yukawa, N. Ito, *Computer Simulation Studies in Condensed Matter Physics XVIII.* D.P. Landau, S.P. Lewis, H.B. Schüttler (Eds.) (Springer, Berlin Heidelberg New York 2006).
28. H. Shiba, S. Yukawa, N. Ito, in preparation.
29. F. Ogushi, *Simulation Study of Heat Conduction Phenomena of Lennard-Jones Particle System.* (Master Thesis, The University of Tokyo, 2005).
30. A. Kamimura, S. Yukawa, N. Ito, J. Phys. Soc. Jpn. **74**, 1071 (2005).
31. A. Kamimura, S. Yukawa, N. Ito, J. Phys. Soc. Jpn. **75**, 024005 (2006).

8

Visualizing Nanodiamond and Nanotubes with AViz

J. Adler, Y. Gershon, T. Mutat, A. Sorkin, E. Warszawski, R. Kalish, and Y. Yaish

Technion-IIT, Haifa, Israel, 32000

Abstract. Carbon atoms form a surprising variety of geometrical structures with a wide range of geometrical, physical and chemical properties. Our AViz atomistic visualization software enables visualization of the results of atomistic simulations of carbon allotropes. Techniques to aid in the perception of the three-dimensional structural specifics of different carbon nanosystems, both alone and in interaction with hydrocarbons are discussed in this chapter.

8.1 Introduction

Visualization is very helpful for atomistic simulations, and the AViz package developed [1–3] by the Computational Physics Group at the Technion is an easy way to implement visualization with true three dimensional quality on simple LINUX boxes. It is useful for both teaching and research applications.

AViz is simple to apply. A file of the x, y and z coordinates of atoms is prepared, either from a simulation or by construction, and then interactively enhanced by adding bonds of varying lengths. The visualization can be updated during a simulation, sliced to show the sample interior, or animated to show time development. We zoom in and out or rotate the sample in order to get a feeling about that third dimension out of a two dimensional screen. In recent years we have been concentrating on extending AViz' applications to general situations where three dimensional visualization can help us deduce what is going on inside samples. Some of these extensions have required new programming but most just require deducing how to utilize features that already exist.

Carbon poses a special challenge to visualization because the different hybridizations lead to different numbers of neighbours, including two (e.g. linear chains or carbyne phase), three (e.g. graphite) and four (e.g. diamond). The projects from which the visualization issues relating to carbon have been extracted and will be described below all have been/will be described in separate publications and presentations together with the physical issues, as well

Springer Proceedings in Physics, Volume 123
Computer Simulation Studies in Condensed-Matter Physics XIX
Eds.: D.P. Landau, S.P. Lewis and H.-B. Schüttler

as qualitative results. Here we concentrate on visualization issues in greater depth than might be appropriate elsewhere. We note that all projects have in common the use of visualization to debug, deduce at which coordinates detailed measurement is needed, show experimental partners what is happening, and finally make presentations.

Our simulations are made with a range of techniques – Molecular Dynamics (MD), Monte Carlo (simulated annealing and simulated tempering), (both with Tersoff/Brenner [4,5] potentials), tight-binding MD (Oxon [6] and Plato [7]), abinitio (Abinit [8]) as suited (and possible). We collaborate extensively with experimentalists at the Technion, and the visualizations form the basis for discussing results with them. In general, our visualizations do not tap directly into the electronic stucture to decide about the hybridization, but rely purely on geometric information. The geometric information of how many nearest neighbours are present largely gives the correct hybridization and certainly enables us to gather useful information because we are, after all, trying to visualize geometric structure.

If carbon only took the two forms of diamond and graphite, that would be enough to make it special. However, as well as carbon being one of the building blocks of organic molecules, pure carbon takes the form of buckyballs, nanotubes etc. And if that were not sufficient, in addition to cubic diamond, there is another, hexagonal solid known as londsdaleite, whose distinction from the cubic form makes most other carbon visualization issues trivial. (The above ode may be sung to the tune of "dayenu" from the Passover seder, if desired.)

We now describe some specific issues. Because we indeed wish to view samples in color and undergoing animation, we refer the reader to the webpage [9] where these are so presented.

8.2 Red, Blue and Yellow

The first steps in attacking carbon specific visualization issues in our Computational Physics group were made by David Segev (Saada) when studying damage in diamond [10]. He also developed our first OpenGL codes for atomistic visualization. The simple idea of using different colors for threefold and fourfold coordinated carbon and in particular of using different colors to indicate bonds of different lengths enabled us to find defects and graphitic structures hiding in 5000 atom samples. We originally used red/blue for threefold and fourfold coordinated atoms, as well as yellow for fourfold coordinated atoms that are neighbours of threefold coordinated ones, and later switched to yellow/blue for three/fourfold coordinated atoms so that the images were understandable in greyscale. More recently some of the graphics was restyled, using colored bonds to join threefold coordinated atoms, leading to a clearer view of graphitic damage in diamond [9].

Differentiating between cubic diamond and graphite is easy if there are large samples of perfectly ordered material. In disordered or amorphous samples this is much less clear and has been an ongoing challenge for us. Recent results from a study [11] of diamond nucleation under high pressure were possible mainly because the atoms were color coded for coordination number and then the diamond nanocrystals became visible within the amorphous matrix.

Possibly the hardest challenge has been differentiating visually between cubic and hexagonal diamond. This is hard even with large ordered samples, because the structures are different only in a subtle way. (It is also hard from an experimental viewpoint.) This is harder when searching for nanocrystals embedded in an amorphous matrix, but the possibility to extract the nanocrystals and then rotate them and compare with reference cubic and hexagonal crystals can result in unambiguous identification of the correct form.

8.3 Transporting, Vibrating and Bending Nanotubes

Static nanotubes can be drawn in many ways, and there are lots of nice pictures on the web. Their vibrations and distortions are less easy to illustrate and visualizing their interactions with other molecules even less so. The issue of transport of hydrocarbons in nanotubes is of interest for applications to the chemical and pharmaceutical industries. The nanotube (which is the flow region) is bounded by two control volumes (CV) with fixed atomic concentrations at the edges to study chemical-gradient driven diffusion. The chemical potential in each CV is fixed by inserting and deleting particles according to the Grand Canonical Monte Carlo (GCMC phase), and the dynamic motion is described by Molecular Dynamics. The particles diffuse through the tube and the parameters of the flow are calculated. In order to show both the tube and the particles passing through it is necessary to find a way to show the tube atoms and bonds being almost transparent so that the diffusing particles can be seen. In order to ensure that they really pass through they are drawn in a different color as they exit, to confirm they have indeed done so. Figure 8.1 shows methane molecules in detail, including bonds for perspective (with Brenner potentials used for both the tube and the methane) entering a tube. For more etensive runs, we used Brenner potentials for short range and Lennard-Jones potentials for long range interactions, and in Fig. 8.2 where the full experiment is shown, for the case of a rigid (6,6) carbon nanotube, of radius 4.07 and length 100 angstroms. The densities of the control volumes in reduced units are CV1 on the right = 0.006, and CV2 on the left = 0.0003. Here we illustrate methane by the larger single spheres, and do not show the bonds of the nanotube explicitly to aid in the transparency. A full characterization of this diffusion will be provided in the future [12].

An experimental group of systems of current interest are nanoscale systems in which there is a coupling between the mechanical and the electrical degrees of freedom. For example, Nano-Electro-Mechanical Systems (NEMS)

Fig. 8.1. Nanotube with methane molecules

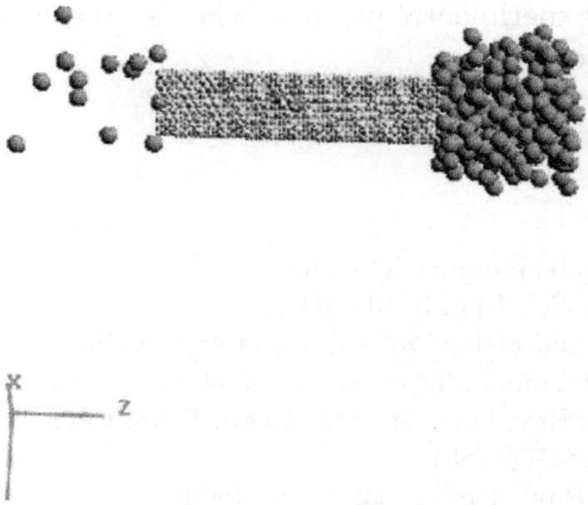

Fig. 8.2. Nanotube with methane molecules diffusing between two control volumes

are being studied in order to realize ultra-sensitive mass resonators, with the aim of achieving the ultimate single molecule detection limit. There are technical limitations to manufacturing smaller nanomechanical beams with conventional photo or e-beam lithography techniques, and degradation is expected in the quality factor as the surface-to-volume ratio increases. An alternative approach is to use carbon nanotubes as the mechanical resonators.

To model this it is necessary to study nanotube vibrations, so we began this by clamping a nanotube at both ends and allowing it to move under a downward force [13]. We studied the best way to visualize the three dimensional tube and found that by indicating the coordinates of each atom as a dot and drawing bonds of a specific length in each case one could clearly see how the tube bends by contacting bonds at the top center and lower ends and lengthening bonds at the top ends and lower center. When the tube vibrates again upward the extended bonds contract.

8.4 Zoom

Another topic of interest is the visualization of different carbon structures such as bucky balls, C240 etc. A two dimenional straight-on picture fails to do justice to their extreme symmetry. Simple rotation of an image does a better job, but successive zooms with a cutoff so that one side of a tube can be viewed at a time (as shown on the talk webpage links [9]) really clarifies the structure.

Acknowledgements

We thank previous and current group members who have collaborated in the AViz project, the authors referenced below who helped us use their software packages and our experimental partners who asked the questions that we use AViz to help answer.

References

1. http://phycomp.technion.ac.il/~aviz
2. J. Adler, Comp. Sci. Eng. **5**, 61 (2003).
3. AViz AtomicVizualization, 2001 Computational Physics Group, Israel Institute of Technology Technion, 32000 Haifa Israel, Geri Wagner, Adham Hashibon.
4. J. Tersoff, Phys. Rev. Lett. **56**, 632 (1986); Phys. Rev. B **37**, 6991 (1988); Phys. Rev. Lett. **61**, 2879 (1988).
5. D.W. Brenner, Phys. Rev. B **42**, 9458 (1990).
6. A.P. Horsfield, A.M. Bratkovsky, D.G. Pettifor, M. Aoki, Phys. Rev. B **53**, 1656, (1996).
7. A.P. Horsfield, Phys. Rev. B **56** 6594 (1997); S. Kenny, A. Horsfield, H. Fujitani, Phys. Rev. B **62** 4899 (2000).
8. http://www.abinit.org/
9. http://phycomp.technion.ac.il/ phr76ja/athens06/talkindex.html
10. D. Saada, J. Adler, R. Kalish, Phys. Rev. B, **59**, 6650 (1999).
11. A. Sorkin, J. Adler, R. Kalish, Phys. Rev B. **70**, 064110 (2004) and in preparation.
12. T. Mutat et al., in preparation.
13. https://phelafel.technion.ac.il/ syanivg/nanotubes/project.htm

9

Molecular Dynamics Simulations for Anisotropic Systems

K.M. Aoki

Institute of Computational Fluid Dynamics aoki@icfd.co.jp

9.1 Introduction

By using conventional molecular dynamics(MD) methods, it is often difficult to properly simulate anisotropic systems. This is because the imaginative boundaries under periodic boundary conditions, i.e., the simulation cell, do not obey the correct dynamics of the system which originate from the dynamics of the molecules inside them. Periodic boundary condition is a clever tool to avoid simulating the whole size and substituting with a part of it. However, in anisotropic systems, periodic boundaries must be used with an appropriate cell dynamics to avoid artifacts induced by them.

9.2 Simulation Cells with Anisotropic Shape

A simple case is when the volume of the system we would like to simulate is not cubic but has an anisotropic shape. If we simulate a system in *an elongated cell* using the Parinello and Rahman (P&R) method [1], the fluctuation of the cell length will differ for the same value of piston mass in all directions (see Fig. 9.1 of [2]) which is a consequence of the P&R method not satisfying the virial theorem. This makes the quality of statistics for a fixed time rage quite different in each direction because of the different period of fluctuation. The original P&R method has been modified to satisfy the virial theorem by introducing a different dynamics in the simulation cell [2]. This makes it possible to simulate even strongly anisotropic systems by a single piston mass.

9.3 Simulating Anisotropic Liquids

Much more difficulty arise when the source of anisotropy is not external (comes for the simulation cell) but also internal as well. Liquid crystal phases are a

Springer Proceedings in Physics, Volume 123
Computer Simulation Studies in Condensed-Matter Physics XIX
Eds.: D.P. Landau, S.P. Lewis and H.-B. Schüttler

state of matter were the physical properties are anisotropic. In lipid membrane, anisotropy and complexity appear in many different scales from the constituent molecules up to the systematic structure as an interface. In such systems, anisotropy in both space and time scale appear as a spontaneous consequence of the dynamics of the molecules inside the simulation cell. In contrast to crystals, smectic liquid crystals and lipid membranes posses elasticity but only in the direction normal to the layers. The layer itself is liquid and only elastic under uniform compressive force. As liquids need a container to avoid spreading out, liquid crystals also need some support in the direction parallel to the layers to maintain the number of smectic layers. The layer thickness compressibility depend on temperature, changing especially drastically near the phase transitions [3]. There is also a large difference in the time scales of fluctuations parallel and perpendicular to the layers in these systems. The temperature dependence of the layer thickness requires the simulation cell to change its shape not matching the periodicity in the fixed cell shape of constant volume or Andersen's constant pressure method [4]. When the cell shape doesn't match the internal periodicity of the system, stress is induced to the system and uniform hydrostatic pressure is not achieved. The absence of elasticity in the direction parallel to the layers results in a large strain fluctuation in this direction, thus leading to the divergence of cell length in the P&R method [1]. In other words, the cell is too flexible in the P&R method to avoid the anisotropic soft matter from spreading out. We need a cell dynamics which can retain the shape to a certain degree but still flexible enough to allow the the anisotropic change in the shape of cell without causing any stress inside the system.

We have achieved this by introducing an anisotropic factor α of the simulation cell in the appropriate Lagrangian of the system [5]. The choice of an appropriate cell variable and cell dynamics (kinetic energy of the simulation cell in the Lagrangian) is the key to get successful simulation results in anisotropic systems.

9.4 Simulations under Constant Surface Tension

Similar problems, such as continuous expansion or contraction of the cell lengths observed in simulating anisotropic liquids, occur in simulating liquid-liquid interfaces. In systems of liquid-liquid interface, due to the interface (and the anisotropy of the composite liquids), there exists large differences in visco-elastic properties depending on the direction (along and normal to the interface) which lead to expansion or contraction of cell lengths (see for instance Fig. 5 in [6]). This is due to the simulation cell dynamics of the methods used where the cell dynamics is described by the equation of motions of each cell lengths. We introduce the anisotropic factor α of the simulation cell along with the surface area to describe the dynamics of the cell [7,8]. The

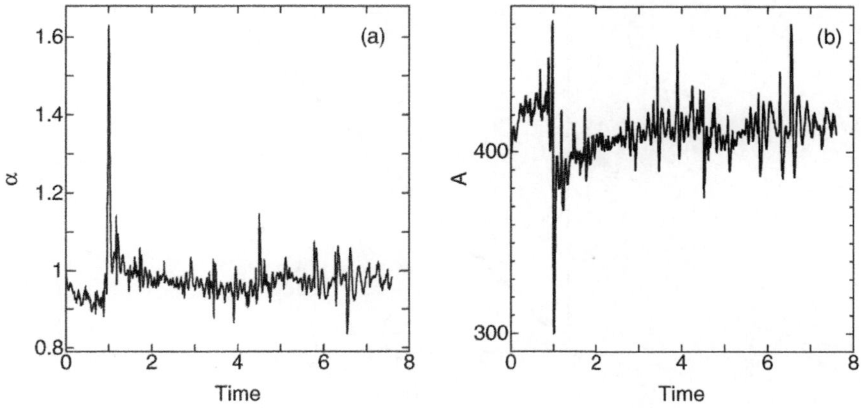

Fig. 9.1. Time evolution of the cell parameters; (**a**) anisotropic factor α and (**b**) surface area A

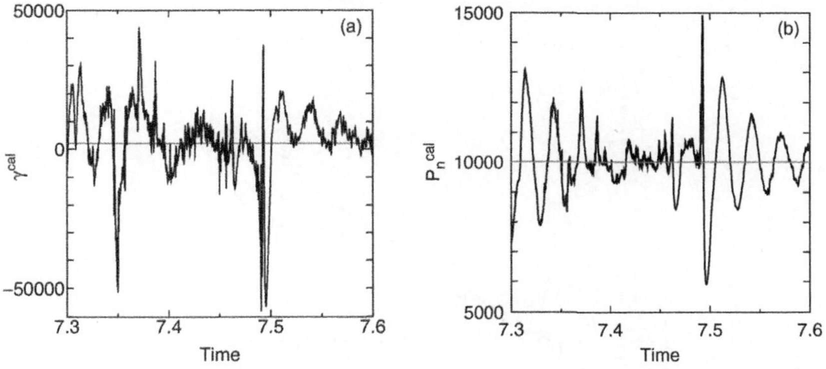

Fig. 9.2. Time evolution of the calculated (**a**) surface tension γ^{cal} and (**b**) normal pressure P_n^{cal}. The values of masss used during this period are $M = 2.0$, $Qt = 1.0$, $Qr = 0.3$

method guarantees that the normal pressure as well as the surface tension balance with the given average values, P_n and γ.

To demonstrate how the method work, a system of soft spherocylinders [9] with length $L = 6$ is calculated. The number of molecules are $N = 2016$, the temperature is $T = 80$ where the effective diameter of the molecules is $d_{\text{eff}} = 0.688$. The given values of the surface tension and the normal pressure are $\gamma = 2.0 \times 10^3$ and $P_n = 1.0 \times 10^4$ respectively. We show the time evolution of the cell parameters α and A in Fig. 9.1. It is shown in Fig. 9.1 that fluctuations of the anisotropic factor α and the surface area A are quite large at some instance. However, there is no sign of continuous expansion or contraction of the cell shape which occur in other methods.

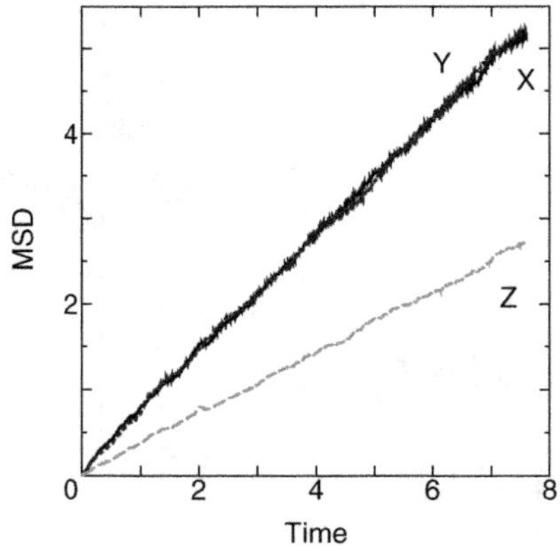

Fig. 9.3. Mean square displacements of the molecules in each direction when $\gamma = 2000$

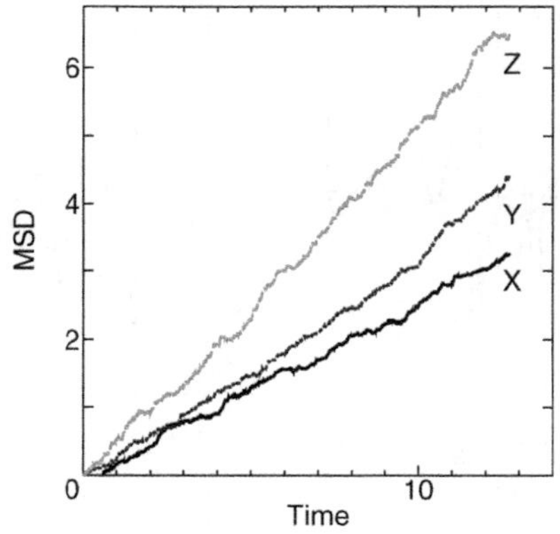

Fig. 9.4. Mean square displacements of the molecules in each direction when the system is under hydrostatic pressure

Next, the calculated values of the surface tension $\gamma_{\rm cal}$ and the normal pressure $P_n^{\rm cal}$ are given in Fig. 9.2. Because of the rigid and strongly anisotropic model we use, the fluctuation of the calculated surface tension is quite large (Fig. 9.2a). However the values of $\gamma_{\rm cal}$ and $P_n^{\rm cal}$ fluctuate around the given values of γ and P_n.

In Fig. 9.3, the mean square displacements(MSD's) in the system under surface tension are shown for each direction;the normal axis to the surface is the z-axis. When there is surface tension, the MSD's in direction parallel to the surface (x & y) are larger than that normal to it. This is in contrast to the MSD's in systems under hydrostatic pressure shown in Fig. 9.4 where the diffusion in z-direction is larger. It should be also noted that the molecules become more aligned when there exists surface tension.

9.5 Concluding Remarks

Simulating anisotropic systems in conventional methods lead to many difficulties. Methods to overcome the difficulties have been proposed in the recent years and have been used to simulate anisotropic liquids. In these methods, the Lagrangian guarantee the desired properties such as virial theorem, hydrostatic pressure or constant surface tension. We hope that the application of these methods to various problems will grow in number in the coming years.

References

1. M. Parinello and A. Rahman, Phys. Rev. Lett. **45**, 1196 (1980).
2. K.M. Aoki, M. Yoneya, H. Yokoyama, J. Chem. Phys **118**, 9926 (2003).
3. J. Yamamoto, K. Okano: Jpn. J. Appl. Phys. **30**, 754 (1991).
4. H.C. Andersen, J. Chem. Phys. **72**, 2384 (1980).
5. K.M. Aoki, M. Yoneya, H. Yokoyama, J. Chem. Phys. **120**, 5576 (2004).
6. Y. Zhang, S.E. Feller, B.R. Brooks, R.W. Pastor, J. Chem. Phys. **103**, 10252 (1995).
7. K.M. Aoki, M. Yoneya, H. Yokoyama, Mol. Cryst. Liq. Cryst. **413**, 109 (2004).
8. K.M. Aoki, M. Yoneya, H. Yokoyama, J. Chem. Phys. **124**, 064705 (2006).
9. K.M. Aoki, T. Akiyama, Mol. Cryst. Liq. Cryst. **262**, 543 (1995); *ibid.* Mol. Simul. **16**, 99 (1996); *ibid.* Mol.Cryst. Liq. Cryst. **299**, 45 (1997); *ibid.* **366**, 117 (2001); D.J. Earl, J. Inytskyi, M.R. Wilson, Mol. Phys. **99**, 1719 (2001); K.M. Aoki, M. Yoneya, H. Yokoyama, Mol. Cryst. Liq. Cryst. **413**, 109 (2004).

10

Event-by-event Simulation of EPR-Bohm Experiments

K. De Raedt[1], K. Keimpema[2], H. De Raedt[2], K. Michielsen[2], and S. Miyashita[3]

[1] Department of Computer Science, University of Groningen, Blauwborgje 3, NL-9747 AC Groningen, The Netherlands
[2] Materials Science Centre, University of Groningen, Nijenborgh 4, NL-9747 AG,Groningen, The Netherlands
[3] Department of Physics, University of Tokyo, Bunkyo-ku, Tokyo 113, Japan

Abstract. We present a computer simulation model that is strictly causal and local in Einstein's sense, does not rely on concepts of quantum theory but can nevertheless reproduce the results of quantum theory for the single-spin expectation values and two-spin correlations in an Einstein-Podolsky-Rosen-Bohm experiment.

10.1 Introduction

Computer simulation is a powerful methodology to model physical phenomena that is complementary to theory and experiment [1]. In this approach, we usually start from the basic equations of physics and employ numerical algorithms to solve these equations. But what if, as in quantum theory, the basic equation that describes the individual events is not known? Last year, also at this workshop, we discussed a simulation method that uses locally-connected networks of processing units with a primitive learning capability to generate events at a rate that agrees with the quantum mechanical probability distribution [2]. The fact that this simulation approach only uses causally local processes raises the question whether can also simulate Einstein-Podolsky-Rosen (EPR) [3] experiments and reproduce the results of quantum theory. This contribution demonstrates that the answer to this question is affirmative.

Quantum mechanical descriptions and experimental realizations of an EPR-Bohm gedanken experiment often adopt the example proposed by Bohm and Aharonov (EPRB) [4, 5]. This model, sketched in Fig. 10.1, considers a source that produces pairs of spin-1/2 particles, prepared in the singlet state $|\mathbf{\Psi}\rangle = (|\uparrow\downarrow\rangle - |\downarrow\uparrow\rangle)/2^{1/2}$. The two particles with opposite spins move in free space and in opposite directions. The spins of the individual particles are measured by means of Stern-Gerlach magnets. After passing a Stern-Gerlach magnet, the particle is detected at either detector $D_{+,i}$ or $D_{-,i}$, where $i = 1, 2$

Springer Proceedings in Physics, Volume 123
Computer Simulation Studies in Condensed-Matter Physics XIX
Eds.: D.P. Landau, S.P. Lewis and H.-B. Schüttler

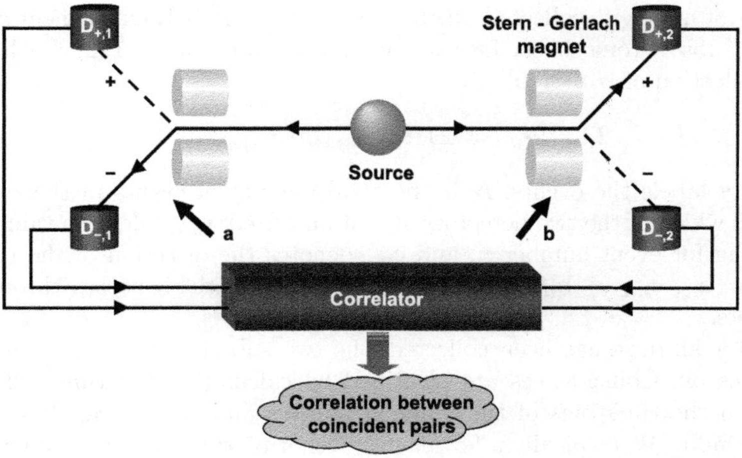

Fig. 10.1. Diagram of the Einstein-Podolsky-Rosen-Bohm experiment

denotes the position of the pair of detectors with respect to the source (see Fig. 10.1). The firing of $D_{+,i}$ ($D_{-,i}$) defines the event at which we assign the value of spin up (+1) (spin down (−1)) to particle i. Representing the direction of magnet 1 (2) by the unit vector **a** (**b**), quantum mechanics yields [6] $\langle\mathbf{\Psi}|\sigma_\mathbf{1}\cdot\mathbf{a}|\mathbf{\Psi}\rangle = \langle\mathbf{\Psi}|\sigma_\mathbf{2}\cdot\mathbf{b}|\mathbf{\Psi}\rangle = 0$ and

$$\langle\mathbf{\Psi}|\sigma_\mathbf{1}\cdot\mathbf{a}\,\sigma_\mathbf{2}\cdot\mathbf{b}|\mathbf{\Psi}\rangle = -\mathbf{a}\cdot\mathbf{b}, \tag{10.1}$$

where $\sigma_i = (\sigma_i^x, \sigma_i^y, \sigma_i^z)$ are the three Pauli spin operators of particle $i = 1, 2$.

Experimentally, each Stern-Gerlach magnet measures the individual spins. Quantum theory itself has nothing to say about these individual assignments (quantum measurement paradox) [7]. The fundamental problem is to explain how individual events, recorded in space-time separated regions under conditions such that the measurement on one particle cannot have a causal effect on the result of the measurement on the other particle (Einstein's local causality criterion), exhibit the correlations (10.1). In this paper, we present a solution of this puzzle.

10.2 Analysis of a Typical EPR-Bohm Experiment

As a first step in solving this puzzle, it is necessary to determine the set of relevant data that is collected in a typical EPRB experiment. Apart from the signals generated by the detectors $D_{+,i}$ and $D_{-,i}$ ($i = 1, 2$), any experimental procedure that registers pairs of particles requires some criterion to decide whether two particles form a pair or not [8]. In EPRB experiments, this decision is taken on the basis of coincidence in time. Thus, a practical criterion for coincidence always involves a coincidence window (see [8, 9]). Therefore, the set of numbers, collected during one run of an EPRB experiment consists of two subsets (one subset for each observation station $i = 1, 2$). Each subset is a collection of triples [9]

$$\Upsilon_i = \{x_{n,i} = \pm 1, t_{n,i}, \mathbf{c}_{n,i} | n = 1, \ldots, N\}, \tag{10.2}$$

where n labels the events, N is the total number of events in the run, $x_{n,i}$ tells us which of the two detectors at station i fired, $t_{n,i}$ holds the value of the time tag for event number n, and $\mathbf{c}_{n,i}$ denotes the direction of the magnets ($\mathbf{c}_{n,1} = \mathbf{a}_n$, $\mathbf{c}_{n,2} = \mathbf{b}_n$) when the nth pair of particles passes through the magnets.

After all data has been collected, the two subsets are analyzed for coincidences [9]. Coincidences are identified by calculating the time differences between the time tags of the different subsets and comparing these with a time window W (typically a few ns [9]). Denoting the number of coincidences between detectors $D_{x,1}$ ($x = \pm 1$) at station 1 and detectors $D_{y,2}$ ($y = \pm 1$) at station 2 by $C_{xy}(\mathbf{a}, \mathbf{b})$, we have

$$C_{xy}(\mathbf{a}, \mathbf{b}) = \sum_{n=1}^{N} \delta_{x, x_{n,1}} \delta_{y, x_{n,2}} \delta_{\mathbf{a}, \mathbf{c}_{n,1}} \delta_{\mathbf{b}, \mathbf{c}_{n,2}} \Theta(W - |t_{n,1} - t_{n,2}|), \tag{10.3}$$

where $\Theta(t)$ is the Heaviside step function and we made a minor abuse of notation by representing the direction of the magnets by discrete labels (which is allowed because in experiment, the number of different directions is necessarily finite, hence representable by integer numbers). Note that the numerator of (10.3) is the number of all detected pairs. The correlation $E(\mathbf{a}, \mathbf{b})$ is given by [5,6]

$$E(\mathbf{a}, \mathbf{b}) = \frac{C_{++}(\mathbf{a}, \mathbf{b}) + C_{--}(\mathbf{a}, \mathbf{b}) - C_{+-}(\mathbf{a}, \mathbf{b}) - C_{-+}(\mathbf{a}, \mathbf{b})}{C_{++}(\mathbf{a}, \mathbf{b}) + C_{--}(\mathbf{a}, \mathbf{b}) + C_{+-}(\mathbf{a}, \mathbf{b}) + C_{-+}(\mathbf{a}, \mathbf{b})}. \tag{10.4}$$

The puzzle to be solved is how to generate the data set $\{\Upsilon_1, \Upsilon_2\}$ under the rather stringent condition that for all events $n = 1, \ldots, N$ and $i = 1, 2$:

$$x_{n,i} = f(\mathbf{c}_{n,i}, \mathbf{S}_{n,i}), \quad t_{n,i} = g(\mathbf{c}_{n,i}, \mathbf{S}_{n,i}), \tag{10.5}$$

such that $E(\mathbf{a}, \mathbf{b}) = -\mathbf{a} \cdot \mathbf{b}$. In (10.5), $\mathbf{S}_{n,i}$ represents the spin of the particle. The functions f and g in (10.5) obey Einstein's criterion of local causality: The values of the measured quantities at station 1 (2) are arithmetically independent of the choice of the settings at station 2 (1), for each individual particle generated by the source.

10.3 Computer Simulation Algorithm

Space limitations prevent us from discussing the motivation that has led us to the following algorithm:

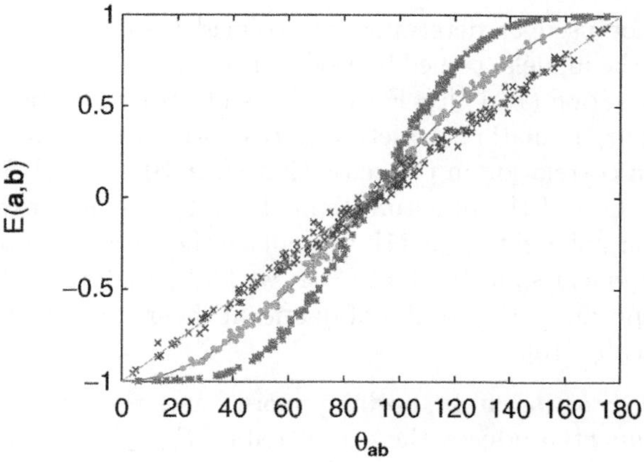

Fig. 10.2. Simulation results for the two-spin correlation for $W/T = \tau/T = 0.00025$, $N = 10^6$, and $M = 200$ randomly chosen values of $\mathbf{a}\cdot\mathbf{b} = \cos\theta_{\mathbf{ab}}$ covering the interval $[-1, +1]$. *Crosses (blue)*: $d = 0$; *bullets (green)*: $d = 3$; *stars (pink)*: $d = 6$; *solid line (red)*: Quantum theory; *dashed line (green)*: $\theta_{\mathbf{ab}}/90 - 1$ (Bell-type model)

1. Specify the number of events N, the time-tag resolution τ/T (the actual value of the time scale T is irrelevant), the time window $W = k\tau$ ($k = 1, 2, \ldots$), and the number M of directions $\mathbf{a}$ and $\mathbf{b}$. Use random numbers to fill the arrays $(\mathbf{a}_1, \ldots, \mathbf{a}_M)$ and $(\mathbf{b}_1, \ldots, \mathbf{b}_M)$ with unit vectors. Set $n = 0$ and $C_{xy}(\mathbf{a}_m, \mathbf{b}_{m'}) = 0$ for all $x, y = \pm 1$ and $m, m' = 1, \ldots, M$.
2. While $n < N$, increment n by one and repeat steps 2 to 6.
3. Use uniform random numbers $-1 \le z_n \le 1$ and $0 \le \phi_n < 2\pi$ to assign the spin $\mathbf{S}_{n,1} = -\mathbf{S}_{n,2} = ((1 - z_n^2)^{1/2}\cos\phi_n, (1 - z_n^2)^{1/2}\sin\phi_n, z_n)$.
4. Use uniform random numbers $1 \le m, m' \le M$ to select the directions $\mathbf{a} = \mathbf{a}_m$ and $\mathbf{b} = \mathbf{b}_{m'}$.
5. The time tag $t_{n,1}$ ($t_{n,2}$) is obtained by generating a uniform random number in the interval $[0, T(1 - (\mathbf{S}_{n,1}\cdot\mathbf{a})^2)^{d/2}]$ ($[0, T(1 - (\mathbf{S}_{n,2}\cdot\mathbf{b})^2)^{d/2}]$) where d is a parameter of the simulation model. Compute $x = \text{sign}(\mathbf{S}_{n,1}\cdot\mathbf{a})$ and $y = \text{sign}(\mathbf{S}_{n,2}\cdot\mathbf{b})$.
6. Apply the coincidence criterion: If $|\lfloor t_{n,1}/\tau\rfloor - \lfloor t_{n,2}/\tau\rfloor| \le k$ the particles form a pair. Then, increment the count $C_{xy}(\mathbf{a}_m, \mathbf{b}_{m'})$. Go back to step 2.
7. After generating N events, we compute the correlations according to (10.4) and plot the results as a function of $\mathbf{a}\cdot\mathbf{b}$.

It is evident that this algorithm satisfies Einstein's criterion of local causality. The use of random numbers to select $\mathbf{S}_{n,i}$ and $t_{n,i}$ is not essential but convenient [10], and merely mimics the apparent unpredictability of the data.

Figure 10.2 shows simulation results for different values of the parameter d. For $d = 3$, our algorithm generates data that agrees with the quantum theory (solid line in Fig. 10.2). In fact, for $d = 3$ we can prove analytically that $\lim_{W\to 0}\lim_{N\to\infty}\lim_{\tau\to 0} E(\mathbf{a}, \mathbf{b}) = -\mathbf{a}\cdot\mathbf{b}$ [10]. For $d = 0$, the time-tag data is not used to determine the coincidences. Then, our model is a realization of the models studied by Bell, hence it cannot reproduce the correct quantum correlation (10.1) [5]. For $d > 3$, as illustrated by the data for $d = 6$ in Fig. 10.2, our simulation model produces correlations that are "stronger" than quantum correlation in the sense that $|E(\mathbf{a},\mathbf{b}) - E(\mathbf{a},\mathbf{b}')| + |E(\mathbf{a}',\mathbf{b}) + E(\mathbf{a}',\mathbf{b}')|$ can exceed the quantum limit $2\sqrt{2}$. To simulate experiments that use the photon polarization [9,11], we replace the three-component spin $\mathbf{S}_{n,1}$ by a two-component spin [10]. For $d = 2$, we find (results not shown) that the simulation reproduces the results of quantum theory, a fact that we can also prove analytically [10].

Summarizing: Starting from nothing more than the observation that an EPRB experiment produces the set of data $\{\Upsilon_1, \Upsilon_2\}$, we have constructed event-based computer simulation models that do not rely on concepts of quantum theory but reproduce the correlation (10.1) that is characteristic for a quantum system in the most entangled state.

References

1. D.P. Landau, K. Binder, *A Guide to Monte Carlo Simulation in Statistical Physics.* (Cambridge University Press, Cambridge 2000).
2. H. De Raedt, K. De Raedt, K. Michielsen, S. Miyashita, Comp. Phys. Comm. (in press); K. De Raedt, H. De Raedt, K. Michielsen, Comp. Phys. Comm. **171**, 19 (2005); K. Michielsen, K. De Raedt, H. De Raedt, J. Comp. Theor. Nanoscience **2**, 227 (2005); H. De Raedt, K. De Raedt, K. Michielsen, J. Phys. Soc. Jpn. Suppl. **76**, 16 (2005); H. De Raedt, K. De Raedt, K. Michielsen, Europhys. Lett. **69**, 861 (2005).
3. A. Einstein, A. Podolsky, N. Rosen, Phys. Rev. **47**, 777 (1935).
4. D. Bohm, Y. Aharonov, Phys. Rev. **108**, 1070 (1957).
5. J.S. Bell, *Speakable and unspeakable in quantum mechanics.* (Cambridge University Press, Cambridge UK 2003).
6. L.E. Ballentine, *Quantum Mechanics: A Modern Development.* (World Scientific, Singapore 2003).
7. D. Home, *Conceptual Foundations of Quantum Physics.* (Plenum Press, New York 1997).
8. J.F. Clauser, M.A. Horne, Phys. Rev. D **10**, 526 (1974).
9. G. Weihs, T. Jennewein, C. Simon, H. Weinfurther, A. Zeilinger, Phys. Rev. Lett. **81**, 5039 (1998).
10. K. De Raedt et al, to be published.
11. A. Aspect, J. Dalibard, G. Roger, Phys. Rev. Lett. **49**, 91 (1982); Phys. Rev. Lett. **49**, 1804 (1982); W. Tittel, J. Brendel, H. Zbinden, N. Gisin, Phys. Rev. Lett. **81**, 3563 (1998).

Part III

Non-Equilibrium and Dynamic Behavior

11

Fisher Waves and the Velocity of Front Propagation in a Two-Species Invasion Model with Preemptive Competition

L. O'Malley[1], B. Kozma[1], G. Korniss[1], Z. Rácz[2], and T. Caraco[3]

[1] Department of Physics, Applied Physics, and Astronomy, Rensselaer Polytechnic Institute, 110 8th Street, Troy, NY 12180-3590, USA
E-mail: omalll@rpi.edu, kozmab@rpi.edu, korniss@rpi.edu

[2] Institute for Theoretical Physics, HAS, Eötvös University, Pázmány sétány 1/a, 1117 Budapest, Hungary
E-mail:racz@general.elte.hu

[3] Department of Biological Sciences, University at Albany, Albany NY 12222, USA
E-mail: caraco@albany.edu

Abstract. We consider an individual-based two-dimensional spatial model with nearest-neighbor preemptive competition to study front propagation between an invader and a resident species. In particular, we investigate the asymptotic front velocity and compare it with mean-field predictions.

11.1 Introduction and Model

The dynamics of propagating fronts are fundamental in the study of the spread of advantageous alleles, species [1], or opinions [2]. Most notably, Fisher [3] and Kolmogorov et al. [4] first addressed the velocity characteristics of a simple front by way of a reaction-diffusion equation [1], which served as a one-dimensional model for the spread of a favorable gene.

Our study of front propagation envisions introduction of an advantageous allele or a competitively superior species through mutation within [5, 6] or geographic dispersal to [7, 8] a resident population, respectively. Introductions occur rarely, and stochastically in both space and time.

We have shown [5–8] that the time evolution of such systems can be well described within the framework of homogeneous nucleation and growth. In particular, in two dimensions, for sufficiently large systems, the typical time scale (lifetime) until competitive exclusion of the weaker allele or species scales as $\tau \sim (Iv^2)^{-1/3}$, where I is the stochastic nucleation rate per unit area of

Springer Proceedings in Physics, Volume 123
Computer Simulation Studies in Condensed-Matter Physics XIX
Eds.: D.P. Landau, S.P. Lewis and H.-B. Schüttler

the successful clusters of the better competitor and v is the asymptotic radial velocity of the corresponding circular fronts. It is, thus, clear that the full understanding of the dependence of the lifetime on the local rates of the systems requires that of the velocity of the front separating the two alleles or species. Therefore, we focus on the velocity characteristics of a two-dimensional two-species model of invasion with preemptive competition, where the invading species has a reproductive advantage over the residents.

Here we study the velocity of the invading fronts of both planar and circular shapes on $L_x \times L_y$ two-dimensional lattices as a function of the reproduction and mortality rates of each species, and compare them with mean-field predictions. Each lattice site can be empty or occupied by the resident or the invader. A lattice site represents the minimal amount of locally available resources which can sustain an individual. Competition for resources is preemptive, and therefore an individual site cannot be taken by either species until its occupant's mortality makes it available. Preemptive competition is typical for plant species competing for common limiting resources [9–13]. The local occupation number at site $\mathbf{x}$ is $n_i(\mathbf{x}) = 0, 1$, $i = 1, 2$, representing the number of resident and invader species, respectively. Due to the excluded volume constraint, $n_1(\mathbf{x})n_2(\mathbf{x}) = 0$, since two species cannot simultaneously occupy the same site. A species can colonize open sites through *local* clonal propagation only. An individual of either species occupying site $\mathbf{x}$ may reproduce only if one or more of its neighboring sites is empty (here we consider nearest-neighbor interactions only).

Our unit time is one Monte Carlo step per site (MCSS) during which $L_x L_y$ sites are chosen randomly. The occupancy (local configuration) of a chosen site is updated based on the following transition rates. When a site is empty, it can become occupied by species i of a neighboring site, at rate $\alpha_i \eta_i(\mathbf{x})$, where α_i is the individual-level reproduction rate and $\eta_i(\mathbf{x}) = (1/4)\sum_{\mathbf{x}' \epsilon \mathrm{nn}(\mathbf{x})} n_i(\mathbf{x}')$ is the density of species i around site $\mathbf{x}$; $\mathrm{nn}(\mathbf{x})$ is the set of nearest neighbors of site $\mathbf{x}$. If a site is occupied by an individual, it can die at rate μ (regardless of the species). We summarize the local transition rules for an arbitrary site $\mathbf{x}$ as

$$0 \xrightarrow{\alpha_1 \eta_1(\mathbf{x})} 1, \quad 0 \xrightarrow{\alpha_2 \eta_2(\mathbf{x})} 2, \quad 1 \xrightarrow{\mu} 0, \quad 2 \xrightarrow{\mu} 0, \tag{11.1}$$

where $0, 1, 2$ indicates whether the site is empty, or occupied by the resident, or invader species, respectively. We study the regime where $\mu < \alpha_1 < \alpha_2$, so that competition between the two alleles drives the dynamics and the invading species has an individual-level reproductive advantage over the resident.

For planar fronts, we consider an $L_x \times L_y$ lattice with periodic boundary conditions along the y direction. The direction of propagation is along the x direction by virtue of the initial condition; we set a flat front separating the invader species from the residents; for simplicity, to the left (right) of the front, all sites are occupied by the invaders (residents). As the simulation begins, a number of individuals die in a few time steps, and in both domains, away from the front, the densities quickly relax to their "quasi-equilibrium" value where the clonal propagation is balanced by mortality. The competition between the two species, hence, takes place in the interfacial region. Throughout the simulation, we keep track of the location of the invading front, by defining the edge as the right-most location of an individual of species 2, for each row of our lattice. The average position of the front is then recorded for each time step, from which one can extract the front velocity. We also studied the velocity of circular fronts on an $L \times L$ lattice, with an initial condition of a sufficiently large central cluster of the invading species (with radius slightly larger than a critical radius [6, 8]), with all other sites occupied by the resident species. We then keep track of the time-dependent global density of the invaders, from which extracted the average radial velocity of the growing circular cluster. Before discussing our simulation results, we first consider those obtained from the mean-field equations of motion.

11.2 Mean-Field Equations and Propagation into an Unstable State

From the above transition rates (11.1) and the underlying master equation, neglecting correlations between the occupation numbers at different sites, for the ensemble-averaged local densities $\rho_i(\mathbf{x}, t) \equiv \langle n_i(\mathbf{x}, t) \rangle$ one obtains

$$\begin{aligned}\rho_i(\mathbf{x}, t+1) - \rho_i(\mathbf{x}, t) = [1 - \rho_1(\mathbf{x}, t) - \rho_2(\mathbf{x}, t)] \frac{\alpha_i}{4} \sum_{\mathbf{x}' \epsilon \mathrm{nn}(\mathbf{x})} \rho_i(\mathbf{x}', t) \\ -\mu\rho_i(\mathbf{x}, t) \,, \end{aligned} \tag{11.2}$$

$i = 1, 2$. For further insight we take a naive continuum limit of (11.2), yielding

$$\begin{aligned}\partial_t \rho_i(\mathbf{x}, t) = &\frac{\alpha_i}{4} [1 - \rho_1(\mathbf{x}, t) - \rho_2(\mathbf{x}, t)] \nabla^2 \rho_i(\mathbf{x}, t) \\ &+ \alpha_i [1 - \rho_1(\mathbf{x}, t) - \rho_2(\mathbf{x}, t)] \rho_i(\mathbf{x}, t) - \mu\rho_i(\mathbf{x}, t) \,. \end{aligned} \tag{11.3}$$

Spatially homogeneous solutions of these equations, (ρ_1^*, ρ_2^*), are $(0, 0)$, $(1 - \mu/\alpha_1, 0)$, and $(0, 1-\mu/\alpha_2)$. For $\mu < \alpha_1 < \alpha_2$, only the $(0, 1-\mu/\alpha_2)$ fixed point is stable. Thus, the motion of the invading front amounts to propagation into an unstable state [3,4,14–16], a phenomenon that has generated a vast amount of literature [17] since the original papers by Fisher [3] and Kolmogorov et al. [4], with applications ranging from reaction-diffusion systems [18,19], population dynamics [1], epidemics [20] or opinion formation in social systems [2]. At the level of the mean-field equations, the front is "pulled" by its leading edge, and for sufficiently sharp initial profiles, the asymptotic velocity v is determined by this infinitesimally small density of invaders intruding into the linearly unstable, resident-dominated regime. Performing standard analysis on (11.3), one obtains the velocity of the "marginally" stable invading fronts [1, 15–17]

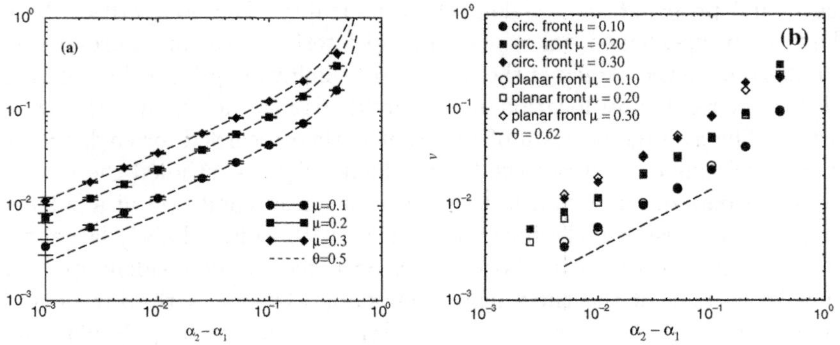

Fig. 11.1. (**a**) Front velocity obtained by numerical iteration of the mean-field equations (11.2), for fixed $\alpha_2 = 0.70$, as a function of the difference of propagation rates, $\alpha_2 - \alpha_1$, for three different values of μ. The dashed curves are the analytic velocities of the "marginally stable" fronts, given by (11.4); the dashed straight line segment corresponds to the slope $\theta = 0.5$. (**b**) Front velocity from Monte Carlo simulations for fixed $\alpha_2 = 0.70$ as in (**a**), with $L_x = 1000$, $L_y = 100$, for planar, and $L_x = L_y = 1000$ for circular fronts, for three values of μ. The dashed straight line segment corresponds to the effective power law with an exponent $\theta \approx 0.62$ for small differences between the local reproduction rates

$$v^* = \frac{\mu}{\alpha_1}\sqrt{\alpha_2(\alpha_2 - \alpha_1)}. \tag{11.4}$$

Thus, for small differences in the local reproduction rates, $v^* \sim (\alpha_2 - \alpha_1)^\theta$ with $\theta = 1/2$. Equation (11.4) reproduces the velocity obtained by numerically iterating the discrete-time discrete-space continuous-density mean-field equations (11.2) remarkable well, as shown in Fig. 11.1a.

11.3 Monte Carlo Results and Discussion

We performed dynamic Monte Carlo simulations using the local rates given by (11.1). In the case of planar fronts, we found that the front velocity is much smaller than that of the mean-field approximation (Fig. 11.1b). Further, for small differences between the local reproduction rates, $v^* \sim (\alpha_2 - \alpha_1)^\theta$ with $\theta \approx 0.62$, an exponent significantly differing from the mean-field scaling (Fig. 11.1). Results from the simulations of the propagation of circular fronts closely match those of the planar fronts. A recent study [20] has found a similar behavior in a discrete two-dimensional stochastic epidemic model.

The discreteness of the individuals (or equivalently, effective density cutoffs in a continuum description) [21–23] and noise [24, 25] have been shown to produce velocity characteristic drastically different from those of the mean-field equations. More precisely, advancing fronts in stochastic individual-based or particle models, which in the mean-field limit converge to a pulled front behavior, are instead "pushed" [17]. That is, the front velocity is determined by the full non-linearity of the frontal region, as opposed to the infinitesimally small leading edge [17]. Our model provides an example for this generic behavior.

Acknowledgments

G.K. is grateful for discussions with Eli Ben-Naim and for the hospitality of CNLS at the Los Alamos National Laboratory, where some of this work was initiated. This research was supported in part by the US NSF through Grant Nos. DEB-0342689 and DMR-0426488. Z.R. has been supported in part by the Hungarian Academy of Sciences through Grant OTKA-T043734.

References

1. J.D. Murray, *Mathematical Biology I and II.* 3rd edn. (Springer, New York, 2002, 2003).
2. E. Ben-Naim, Europhys. Lett. **69**, 671 (2005).
3. R.A. Fisher, Annals of Eugenics **7**, 355 (1937).
4. A.N. Kolmogorov, I. Petrovsky, N. Piskounov, Moscow Univ. Bull. Math. **1**, 1 (1937).
5. J.A. Yasi, G. Korniss, T. Caraco, in *Computer Simulation Studies in Condensed Matter Physics XVIII.* Ed. by D.P. Landau, S.P. Lewis, and H.-B. Schüttler, Springer Proceedings in Physics (Springer, Berlin Heidelberg New York 2006) in press.
6. L. O'Malley, J. Basham, J.A. Yasi, G. Korniss, A. Allstadt, T. Caraco, submitted to Theor. Popul. Biol. (preprint, 2005); arXiv:q-bio/0602023
7. G. Korniss, T. Caraco, J. Theor. Biol. **233**, 137 (2005).
8. L. O'Malley, A. Allstadt, G. Korniss, T. Caraco, in *Fluctuations and Noise in Biological, Biophysical, and Biomedical Systems III.* Ed. by N.G. Stocks, D. Abbott, R.P. Morse. Proceedings of SPIE Vol. 5841 (SPIE, Bellingham, WA, 2005) pp. 117–124.
9. J.B. Shurin, P. Amarasekare, J.M. Chase, R.D. Holt, M.F. Hoopes, M.A. Leibold, J. Theor. Biol. **227**, 359 (2004).
10. P. Amarasekare, Ecol. Lett. **6**, 1109 (2003).
11. D.W. Yu, H.B. Wilson, Am. Nat. **158**, 49 (2001).
12. D.E. Taneyhill, Ecol. Monogr. **70**, 495 (2000).
13. B. Oborny, G. Meszéna, G. Szabó, Oikos **109**, 291 (2005).
14. D.G. Aronson, H.F. Weinberger, Adv. Math. **30**, 33 (1978).
15. G. Dee, J.S. Langer, Phys. Rev. Lett. **50**, 383 (1983).
16. W. van Saarloos, Phys. Rev. Lett. **58**, 2571 (1987).
17. W. van Saarloos, Phys. Rep. **386**, 29 (2003).
18. J. Riordan, C.R. Doering, D. ben-Avraham, Phys. Rev. Lett. **75**, 565 (1995).
19. D. ben-Avraham, Phys. Lett. **247**, 53 (1998).
20. C.P. Warren, G. Mikus, E. Somfai, L.M. Sander, Phys. Rev. E **63**, 056103 (2001).

21. E. Brunet, B. Derrida, Phys. Rev. E **56**, 2597 (1997).
22. D.A. Kessler, Z. Ner, L.M. Sander, Phys. Rev. E **58**, 107 (1998).
23. D.A. Kessler, H. Levine, Nature **394**, 556 (1998).
24. C.R. Doering, C. Mueller, P. Smereka, Physica A **325**, 243 (2003).
25. J.G. Conlon, C.R. Doering, J. Stat. Phys. **120**, 421 (2005).

12

Dynamics and Thermal Structure of Gas-Liquid Phase Interface

F. Ogushi, S. Yukawa, and N. Ito

Department of Applied Physics, School of Engineering, The University of Tokyo, 7-3-1 Hongo, Bunkyo-ku, Tokyo 113-8656

Abstract. The dynamics and the thermal structure of the gas-liquid phase interface is studied using nonequilibrium molecular dynamics simulation in Lennard-Jones particle system of three-dimension. The initial configuration of the system is in equilibrium state where the temperature and density keeps definite value. At $t = 0$, temperature gradient charges on the system box in x-direction and the system starts phase separation close to the steady state. From unstationary state, a gap of temperature profile $\Delta T(t)$ between gas phase and liquid phase is exposed and this interface profile is settled to a definite value ΔT_0 with $1/\sqrt{t}$. This ΔT_0 is corresponding to a gap of the temperature profile in the steady state. This result suggests that the thermal resistance in the gas-liquid phase interface is shown in this scale.

12.1 Introduction

Heat conduction is one of the most fundamental issues in physics and technology. Although its study has a long history dating back to Fourier's day in the nineteenth century. In macroscopic scale, the heat conduction is well described by the Fourier's law of heat conduction but the theoretical description based on microscopic dynamics is still unsatisfactory. To understand the heat conduction of nonequilibrium steady state, theoretical description of the structure of the gas-liquid phase interface is one of the biggest challenge. The gas-liquid interface is relevant to the nucleation [1]. Volumer and Becker give a simple theory to explain the nucleation that estimate the critical nucleus using the free enrgy [1]. It is called the classical nucleation theory (CNT) and it is a simple theory to compare theoretical predictions with experimental results. The CNT dose not consider the finite thickness of the interface and this theory dose not explain the bubble nucleation. Oxtoby et al. modify the CNT using the density functional method (DF) [2]. This theory gives diffusive interfaces and it predicts nucleation rates that are more close to the experiment results than the CNT. Thus, the structure of the gas-liquid phase interface has clutial

Springer Proceedings in Physics, Volume 123
Computer Simulation Studies in Condensed-Matter Physics XIX
Eds.: D.P. Landau, S.P. Lewis and H.-B. Schüttler

importance in the bubbule nucleation. As one example of the heat conduction phenomenon in the interface, the heat resistance is known, for example Kapitza resistance [3] that a gap of the temperature profile is shown in the interface and the heat conducitivity in the area becomes very small. Kapitza resistance is the phenomeno in the ^{3}He-solid interface and this is explained using phonon scatterling [4, 5] but this explanation is insufficient for general thermal resistance. The heat conduction phenomena of multi phase system is studied using computer simulation. Using nonequilibrium molecular dynamics simulation, the heat conduction of the multi pahse system and the structure of the pahse interface are studied [6, 7]. About the density profile of gas-liquid phase interface, it has an asymmetric structure and it shows good agreement with an asymmetric free enerfy density model with dobule-well form. The temperature shows linear profile. The temperature gradient is different in each phase side but the detail structure in the interface dose not clear.

To study of the heat conduction of the gas-liquid phasae interface, we need study about more detail thermal structure of the interface. But the simulation of the gas-liquid coexisting state takes very high cost. Thus, we study the dynamics and thermal structure of the gas-liquid phase interface from unstationary state [7].

12.2 Model and Simulation

Lennard-Jones (12-6) potential is described as

$$\phi(r_{ij}) = \begin{cases} 4\epsilon\left\{\left(\frac{\sigma}{r_{ij}}\right)^{12} - \left(\frac{\sigma}{r_{ij}}\right)^{6}\right\} & (r \leq r_c) \\ 0 & (r > r_c), \end{cases}$$

where r_{ij} and σ denote a distance between particle i and j and a diameter of each particle, respectively. We set σ, ϵ, and the mass of the particle m to be equal to one.

The geometry of the simulation system is a three-dimensional rectangular paralellpiped box $L_x \times L_y \times L_z$ where L_i is the system size length in i-direction $(i = x, y, z)$. The boundary condition along y- and z-direction is periodic to reducing undesired boundary effects on the bulk properties. Along x-direction, there are elastic walls on the both ends of the system box. The particle can go out across the end of the system box but a potential force according to the distance from the end of the system box works on the particle out of the box and takes the particle back to the system box.

To give a temperature gradient, both end regions in x-direction are contact with Nosé-Hoover heat bathes with different temperature T_H and T_L ($T_H = 1.6$, $T_L = 0.8$). In the Nosé-Hoover heat bath, the temperature difference of the particle and the heat bath we set T_{HB} is controlled as a friction force. The equation of motion in the heat bath is

$$m\frac{d^2r}{dt^2} = F - \zeta m\frac{dr}{dt},$$

$$\frac{d\zeta}{dt} = \frac{2(K - K_{\mathrm{HB}})}{Q},$$

where r and ζ, K, K_{HB}, Q are the position of a particle, heat bath variable, the kinetic energy of the particle, the kinetic energy of the heat bath and the time constant, respectively.

The particles are updated using the Verlet method and the Nosé-Hoover heat bath is calculated with the Euler method. By those two methods, time step is fixed and set to be 10^{-4}. The density of the particle ρ is set to be about 0.3 and the system size is set to be $L_x \times L_y \times L_z = 50 \times 6 \times 6$. the length of heat bath area in x-direction is dx_{HB} is set to be five.

In order to study the profile of the temperature and the density, we slice the system box with a fixed width dx_{cell} into a local cell. Local equilibrium is based on the difference between the thermal scale of each particle and the macroscopic mass flow scale. This is unclear in the microscopic world. Murakami et al. studied about the local equilibrium for nonequilibrium steady state [6]. They confirmed that the local velocity distribution has an approximately Maxwellian form. The local equilibrium is thought to be valid and the local temperature is defined in the same way as that in the equilibrium state. For the Lennard-Jones system, we also use the following definitions: The local temperature $T(x)$ and density $\rho(x)$ are defined as

$$T(x) = m\frac{\langle \boldsymbol{p_i}^2 \rangle_{\mathrm{cell}}}{3k_{\mathrm{B}}}, \tag{12.1}$$

$$\rho(x) = \frac{\langle n \rangle_{\mathrm{cell}}}{V}, \tag{12.2}$$

where the $\langle \ldots \rangle_{\mathrm{cell}}$ is a long time average in each cell. The n is a number of particles in the cell and V is a volumn of the cell $V = L_y \times L_z \times dx$. In this article, we measure the temperature, the density and the length in units of ϵ/k_{B}, σ^{-2} and σ where k_{B} is Boltzmann constant and it is set to be one.

12.3 Result

At the initial configulation, the system is in a equilibrium state where the temperature and the density are constant ($T(x) = 1.6$, $\rho(x) = 0.3$). At time $t = 0$, a temperature gradient is charged on the system box in x-direction and the heat flux drives the system into gas-liquid coexisting phase. Finally, the system reaches a steady state and there is a phase interface steadily in the system About the phase interface structure of the steady state, show the refarence [7]. We study the dynamics of the interface using the temperature and the density profile at unstationary state.

12.3.1 Dynamics of Density Profile

At time $t = 0$, the system begins the phase separation. The liquid phase begins to grow up from T_L side and the interface of the density profile closes to the position at the steady state (shown in Fig. 12.1). At the first stage ($t < 300$), the liquid pahse grows up with $\sqrt{t}$ and this behavior is consistent with the result in an infinite system from the diffusion equation. The position at $\rho(x) = 0.4$, this is the typical value, in the interface, is adjusted to the position $x(t)$ of the interface at time t. After this process ($t > 300$), the position of the interface is settled to a definite value x_0 with $1/\sqrt{t}$. The x_0 is corresponding to the equilibrium position in the steady state. The distance from $x(t)$ to x_0 is shown in Fig. 12.2.

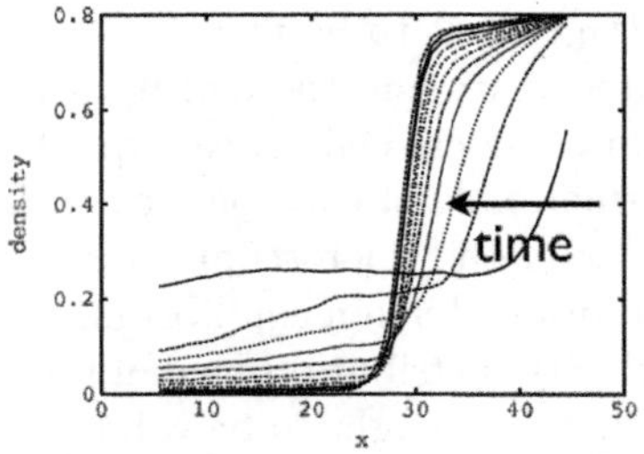

Fig. 12.1. Density profile $\rho(t)$

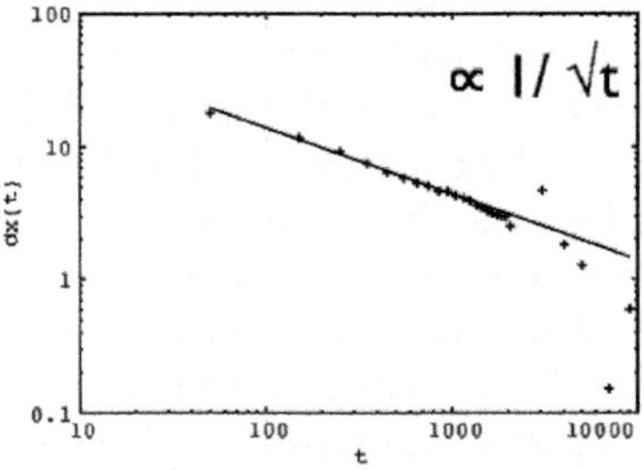

Fig. 12.2. Distance from the equilibrium position of the interface

12.3.2 Dynamics of Temperature Profile

At the initial configulation, the system is in an equilibrium state and the temperature is constant $T(x) = 1.6$. At time $t = 0$, the system begins a phase separation into the gas-liquid coexisting state. The temperature gradient loose in liquid phase side and becomes tight in gas phase side as it approaches the steady state. In the steady state, Fourier type heat conduction is realized and the different temperature gradient is shown in each phase side. The temperature profile $T(x)$ is shown in Fig. 12.3. At the initial stage, the density of T_L side rises rapidly and mass flow is caused from T_H side to T_L side. Therefore, extra energy collects and there is little difference in the temperature between the gas phase side and the liquid phase side. After this, there is a gap of the temperature profile $\Delta T(t)$ is shown between gas phase side and liquid phase side. The behavior of $\Delta T(t)$ is shown in Fig. 12.4. The value of $\Delta T(t)$ is settled to a definite value ΔT_0 that is corresponding to the value of a gap of $T(x)$ in the steady state. This result suggests that the heat resistance, a gap of temperature profile in the interface, is shown in the gas-liquid interface in this scale.

12.4 Summary

Using nonequilibrium molecular dynamics simulation, the dynamics and thermal structure of gas-liquid phase interface is studied from unstationary state.

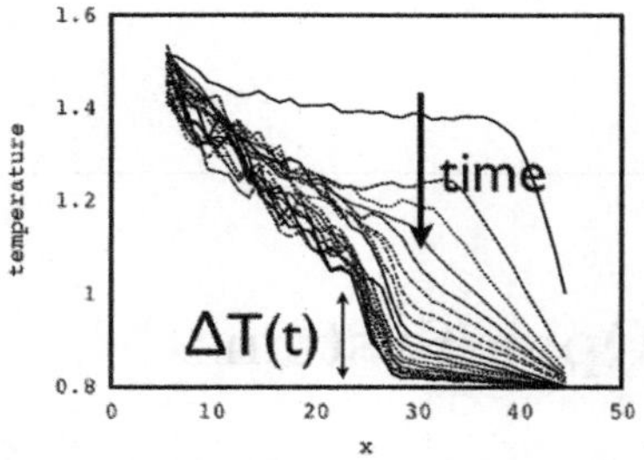

Fig. 12.3. Temperature profile $T(t)$

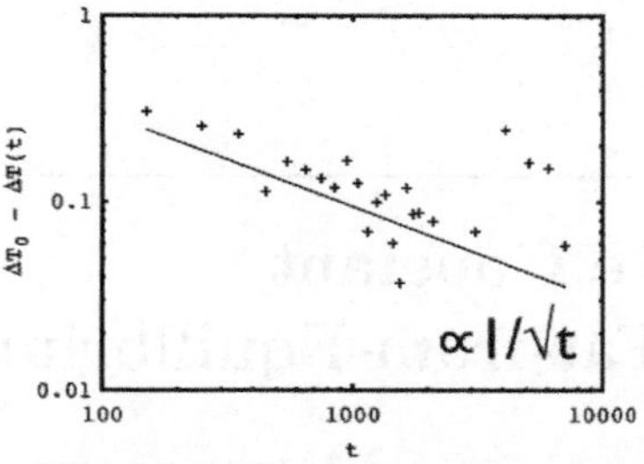

Fig. 12.4. Time dependence of $\Delta T_0 - \Delta T(t)$

The geometry of the system box is a rectangular pallarele piped box in three-dimension. At the initial configuration, the system is in an equilibrium state that the temperature and the density are constant. At time $t = 0$, the temperature gradient is charged on the system box and the system begins to separate into a gas-liquid coexisting state.

After time $t = 0$, the liquid phase begins to grow up from the low temperature heat bath side. At the first stage ($t < 300$), the thickness of liquid pahse $l(t)$ is proportional to $\sqrt{t}$ and this behavior corresponds to the result in a infinite size system from the diffusion equation. After this, the interface position of density profile is settled to the equilibrium position in a steady state with $1/\sqrt{t}$.

About the temperature profile, it shows a linearity form and the temperature gradient is different in each phase side. There is a gap of temperature profile $\Delta T(t)$ between gas phase side and liquid phase side. The value of $\Delta T(t)$ is settled to a definite value ΔT_0 with $1/\sqrt{(t)}$. The differece of the internel energu between both ends of the interface almost corresponds to the value of $\Delta T(t)$. This suggests that the heat resistance is shown in a gas-liquid phase interface in this scale.

References

1. M. Volmer, H.Z. Flood, Z. Phys. Chem. A **190**, 273 (1934); R. Becker, W. Döring, Ann. Phys. **24**, 719 (1935); J. Frenkel, *Kinetic Theory of Liquids.* (Theodre Steinkopff, Dresden, 1939).
2. F.F. Abraham, *Homogeneous Nucleation Theory.* (Academic Press, New York, 1979); A. Laaksonen, V. Talanquer, D.W. Oxtoby, Annu. Rev. Phys. Chem. **46**, 489 (1995); J.M. Howe, Philos. Mag. A **74**, 761 (1996).
3. P.L. Kapitza, J. Phys. (MOSCOW) **4**, 181 (1941).
4. E.T. Swartz, R.O. Pohl, Reviews of Modern Physics **61**, 605 (1989).
5. D.G. Cahil, W.K. Ford, K.E. Goodson, G.D. Mahan, A. Majumdar, H.J. Maris, R. Merlin, S.R. Philpot, J. App. Phys. **93**, 793 (2003).
6. T. Murakami, T. Shimada, S. Yukawa, N. Ito, J. Phys. Soc. Jpn. **72**, 1049 (2003).
7. F. Ogushi, S. Yukawa, N. Ito, in printing in proc. of the 18th Anual Workshop on XVIII (2005).

13

Rate Constant in Far-from-Equilibrium Open Systems

A. Kamimura, S. Yukawa, and N. Ito

Department of Applied Physics, School of Engineering, The University of Tokyo, 7-3-1 Hongo, Bunkyo-ku, Tokyo, 113-8656, Japan

Abstract. As a first step to study reaction dynamics in far-from-equilibrium open systems, we propose a stochastic model in which replicating reactions with catalysis progress depending on the external flow of energy resources J. This model exhibits the Arrhenius-type reaction; furthermore it produces non-Arrhenius reaction that is power-law reaction rate with regard to reaction energy. This behavior is explained using dynamics of J, especially the power law is explained by the conservation of J.

13.1 Introduction

Chemical reaction systems have intrinsic fluctuations in composition since collisions between chemicals may change their identities. An isolated system attains an equilibrium point along with the second law of thermodynamics, and, in simple reacting systems, dynamics of the systems are written by rate equations. Reaction rate is proportional to the multiplier of concentrations of reactants and its proportional constant κ is referred to as a rate constant. A basic relation between rate constant and temperature T is commonly known as the Arrhenius equation [1], $\kappa \propto \exp(-E/k_{\mathrm{B}}T)$ where k_{B} denotes the Boltzmann constant. The activation energy E is interpreted as a minimum energy to initiate a reaction.

Reacting systems show a great variety of nonequilibrium behavior. Typical example is a biological system. The activity of biological systems can be interpreted as a highly cooperative machinery which is composed by various biochemical reactions. Such systems are characterized as open systems in far-from-equilibrium states. A study of different types of order in such states is pioneered by Prigogine [2] with regard to irreversible thermodynamics. The behavior of reacting system in far-from-equilibrium with various kinds of chemicals have been studied from phenomenological point of view especially in the area of chemical engineering such as combustion [3]. Further, with regard to the origin of life, kinetic approaches to answer a problem that how life continues to replicate precise copies under accidental errors are studied such as hypercycle by Eigen and Schuster [4].

Springer Proceedings in Physics, Volume 123
Computer Simulation Studies in Condensed-Matter Physics XIX
Eds.: D.P. Landau, S.P. Lewis and H.-B. Schüttler

However, it is not well developed from microscopic point of view with regard to statistical mechanics. Especially, the manner in which reactions progress in far-from-equilibrium open systems is not generally understood. As a first step to study the reaction dynamics, we propose a homogeneous model in which reactions progress depending on the external flow of energy resources.

13.2 Models with Energy Flow

We consider a homogeneous system wherein rate equations hold. Because rate constants are determined by the dynamics of molecules, we do not fix rate constant. Instead, we introduce a new quantity J which is considered to be a resource for the occurrence of reactions such as substances, light, the number of ATP and so on. We refer to J as energy in this paper. It is assumed that J is flowing into the system constantly by ΔJ, irrespective of the occurrence of reactions. We also determine E for each reaction which is necessary to progress the reaction. We progress reactions when the system has enough resources J for reactions. These dynamics determine rates of reactions.

It would be realistic that energy resource is distributed to every molecule, and every molecule react according to its own energy. However, the system is homogeneous and we focus on a steady-state feature. Thus dynamics from the point of every molecule and from the whole system would be essentially identical. Therefore we assume that J is a parameter of the whole system.

As explained in the following subsection, we adopt replicating systems with catalytic chemicals as simple biological reacting models [5, 6].

We consider that molecules are placed in a container. We introduce discrete simulation steps and take the following procedures in every step. We choose a pair of molecules i and j randomly from the container. If there is a reaction path between the two molecules, reaction may happen, say i is catalyzed by j. If J at the moment is greater than reaction energy of the reaction E, reaction occurs. Then, another molecule of chemical i is added into the container and E is subtracted from J. Otherwise, we only put the two molecules back. We add ΔJ to J, irrespective of the occurrence of reactions. We include division events as follows. If the total number of molecules become a threshold $2N$, we remove randomly around N molecules from the system.

13.2.1 Mutually Catalyzing Chemicals

We consider the case that tw mutually catalytic chemicals X and Y are replicating as follows [7],

$$X + Y \to 2X + Y\,, \quad 2Y + X\,. \tag{13.1}$$

Here, either of reactions occurs due to which chemical is considered to be a catalyst. We consider that j of the two chosen molecules explained above is a catalyst in our simulations. We assume that reaction energy of replicating Y,

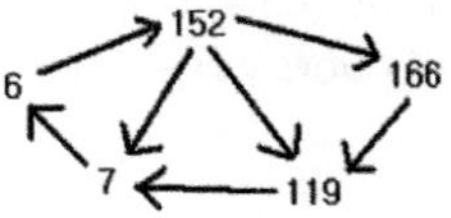

Fig. 13.1. Survived reaction cycle. Parameters are fixed as $N = 8000$, $s = 200$, $\rho = 0.1$, $E_{\mathrm{max}} = 30$ and $\Delta J = 1$. *Numbers* denote chemicals and an *arrow* from X_{i} to X_{j} denotes that X_{j} is catalyzed by X_{i}. The reaction energies of the survived chemicals are $(E_6, E_7, E_{152}, E_{119}, E_{166})$=(11.45, 6.31, 2.65, 7.37, 0.18). Chemicals 6, 7 and 152 compose a *core cycle* and 119 and 166 are peripheral but catalyzed by the *core cycle*

E_{y} is greater than that of X, E_{x}. Details are explained in [7]. The model shows an Arrhenius equation $\gamma_{\mathrm{y}} \sim \exp(-E_{\mathrm{y}}/\alpha\Delta J^2)$ in the range of small ΔJ. Further, it exhibits a non-Arrhenius reaction that is power-law dependence $E_{\mathrm{y}}^{-1/2}$ as ΔJ is fixed to be a greater value. We also find that power changes from $-1/2$ to -1 due to the digitality of molecules. This behavior is explained using dynamics of J. Asymmetric random walk model of J explains the Arrhenius equation and conservation of J results in the power-law dependence.

13.2.2 Catalytic Networks

We consider the case of replicating system with many chemicals in this subsection. The container includes s kinds of chemicals X_{i} $(i = 1, 2, .., s)$ at initial step and they react as follows,

$$X_{\mathrm{i}} + X_{\mathrm{j}} \rightarrow 2X_{\mathrm{i}} + X_{\mathrm{j}} \tag{13.2}$$

according to a randomly chosen catalytic network. Density of path reaction for each chemical is fixed to be ρ, i.e., every chemical has ρs catalysts on average. We neglect reverse reactions and exclude mutually catalytic reactions. We determine reaction energies $E_{\mathrm{i}} \in [0, E_{\mathrm{max}}]$ for chemicals X_{i} randomly.

By simulating this reacting models in some parameters, it is observed that several chemicals are survived and replicating steadily. It is also observed that the survived chemicals composes a reaction cycle as shown in Fig. 13.1. It is noteworthy that the survived chemicals can change even if parameters are the same because they also depends on seeds of random numbers. We will not discuss the survival and examine the properties of survived cycle. As shown in Fig. 13.2, this model shows log-normal distributions in fluctuations of the number of molecules, which is one of characteristic features of biological systems [6]. Analysis of rate constants of the survived chemicals can be carried out using asymmetric random walk model of J in a similar way of the model in Sect. 13.2.1.

In particular, we explain how the rate constant depends on the reaction energy in case that the system should possess a "heavy" chemical which means the chemical needs a great energy E to reproduce. As a simplest case in our model, the cycle is composed by three cyclically catalytic chemicals, say A, B, and C. We refer to the "heavy" chemical as A. Further, we assume that E_{B} and E_{C} are of the order of ΔJ so that rate constants of B and C become one.

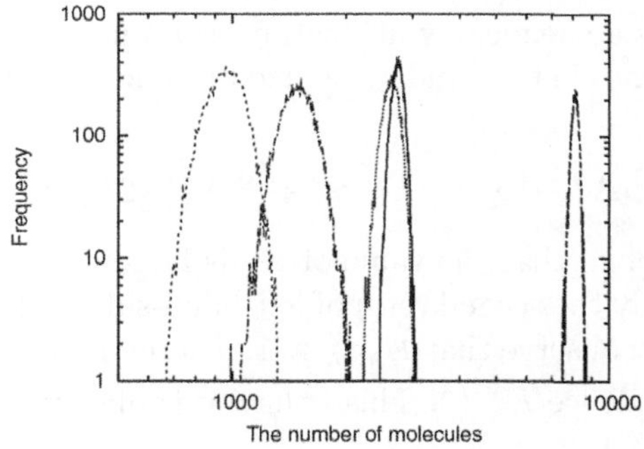

Fig. 13.2. The number distributions of molecules corresponding to the network in Fig. 13.1 at the division events. From *right* to *left*, the chemicals shown are 7, 6, 152, 166 and 119. The two peripheral chemicals show parabolic shapes, that are log-normal distributions

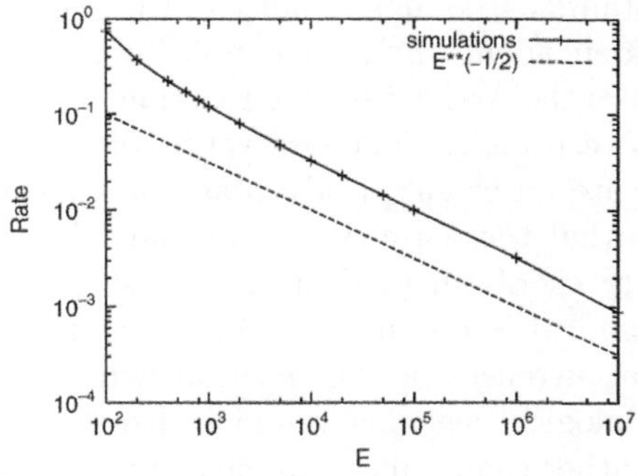

Fig. 13.3. Rate constant of the "heavy" chemical A vs. E_{A}. The slope $E^{-1/2}$ is also drawn

The behavior is explained by the macroscopic relations of the system. In a steady state, the following relations hold in the system: the conservation of energy in each step,

$$\Delta J = E_{\mathrm{A}}\Delta N_{\mathrm{A}} + E_{\mathrm{B}}\Delta N_{\mathrm{B}} + E_{\mathrm{C}}\Delta N_{\mathrm{C}}, \tag{13.3}$$

the condition of steady state about the number

$$\frac{N_{\mathrm{i}} + \Delta N_{\mathrm{i}}}{N_{\mathrm{t}} + \Delta N_{\mathrm{t}}} = \frac{N_{\mathrm{i}}}{N_{\mathrm{t}}}, \tag{13.4}$$

where Δ, i and N_{t} denotes the expecting increase in one step, index of chemicals, and the total number of molecules, respectively. Furthermore, ΔN_{i} is written as

$$\Delta N_{\mathrm{i}} = \frac{N_{\mathrm{i}}}{N_{\mathrm{t}}}\frac{N_{(\mathrm{i})}}{N_{\mathrm{t}}}\gamma_{\mathrm{i}}, \tag{13.5}$$

where (i) indicates a chemical which catalyzes i, and $\gamma_{\rm i}$ is the rate constant.

When we assume that $\gamma_{\rm B}$ and $\gamma_{\rm C}$ is fixed to one, we obtain from (13.3) to (13.5)

$$(\Delta J - E_{\rm B})\gamma_{\rm A}^3 + (5\Delta J - 2E_{\rm A} - E_{\rm B})\gamma_{\rm A}^2 + (8\Delta J - 2E_{\rm C})\gamma_{\rm A} + 4\Delta J = 0. \quad (13.6)$$

If we remind ourselves that the value of $E_{\rm A}$ is large compared with the other parameters and only the second term of left-hand-side of the equation includes the $E_{\rm A}$, we roughly observe that $E_{\rm A}\gamma_{\rm A}^2 \approx 1$. Therefore, the rate constant have a power-law dependence $E_{\rm A}^{-1/2}$. This behavior is observed in our simulations as shown in Fig. 13.3.

13.3 Summary

We have studied reacting models with external flow of energy J. The models determine rate constants based on a balance of the energy flow. These models exhibits power-law dependence in a nonequilibrium steady state. This is essentially different from the Arrhenius type reaction. The behavior is explained using dynamics of J, especially the conservation of J.

This result indicate an essential advantage of some far-from-equilibrium states in a situation that there are essential chemicals to sustain the system, which is Y or A in terms of our models. In such situations, the system must "wait" until the chemical is obtained, and our result of the power-law rate implies significant occurrences of the reaction which is hardly occurred by heat excitations. Biological systems would utilize this advantage to attain high energy states rather than only using heat fluctuations.

Acknowledgement

This work is partly supported by the Japan Society for the Promotion of Science (No. 15607003).

References

1. S. Arrhenius, Z. Phys. Chem. **4**, 226 (1889).
2. G. Nicolis, I. Prigogine, *Self-Organization in Nonequilibrium Systems.* (John Wiley & Sons, 1977).
3. See, e.g., J. Warnatz, U. Maas, R.W. Dibble, *Combustion.* 3rd edn. (Springer, Berlin Heidelberg New York, 2001).
4. M. Eigen, P. Schuster, *The Hypercycle.* (Springer, Berlin Heidelberg New York 1979).
5. K. Kaneko, T. Yomo, J. Theor. Biol. **214**, 563 (2002).
6. K. Kaneko: Phys. Rev. E **68**, 031909 (2003).
7. A. Kamimura, S. Yukawa, N. Ito, J. Phys. Soc. Jpn. **74**, 1071 (2005); J. Phys. Soc. Jpn. **75**, 024004 (2006).

14

First-order Reversal Curve Analysis of Kinetic Monte Carlo Simulations of First- and Second-order Phase Transitions

I.A. Hamad[1,2], D. Robb[2], and P.A. Rikvold[1,2,3]

[1] Center for Materials Research and Technology and Department of Physics, Florida State University, Tallahassee, FL 32306-4350, USA
[2] School of Computational Science, Florida State University, Tallahassee, FL 32306-4120, USA
[3] National High Magnetic Field Laboratory, Tallahassee, FL 32310, USA
hamad@scs.fsu.edu, robb@scs.fsu.edu, rikvold@scs.fsu.edu

Abstract. The dynamics near first- and second-order phase transitions in a two-dimensional lattice-gas model are compared using the first-order reversal curve (FORC) method. The FORC diagram of a first-order transition is characterized by a negative region separating two positive regions, reflecting a competition between the time-varying electrochemical potential and the tendency of the system to phase order.

14.1 Introduction

The first-order reversal curve (FORC) method was conceived [1] in connection with the of magnetic systems, as a means to determine the Preisach distribution. Since its initial application to magnetic recording media [2], the FORC method has been applied to a variety of magnetic systems, ranging from magnetic nanostructures to geomagnetic compounds, undergoing *rate-independent* (i.e., very slow) magnetization reversal. Recently there have also been several FORC studies of *rate-dependent* reversal [3–5].

Here we apply FORC analysis to rate-dependent reversal in lattice-gas models of electrochemical deposition. Specifically, we study a two-dimensional lattice-gas model with attractive nearest-neighbor interactions, being driven across its first-order phase transition by a time-varying electrochemical potential, as well as a lattice-gas model with repulsive lateral interactions and nearest-neighbor exclusion, being driven across its second-order transition.

For the lattice-gas system, the FORC method consists of saturating the adsorbate coverage in a strong positive electrochemical potential, and, in each case starting from saturation, decreasing the potential to a series of progressively more negative 'reversal potentials' $\bar{\mu}_r$, and subsequently increasing the

Springer Proceedings in Physics, Volume 123
Computer Simulation Studies in Condensed-Matter Physics XIX
Eds.: D.P. Landau, S.P. Lewis and H.-B. Schüttler

potential back to the saturating value [2]. This produces a family of FORCs $\theta(\bar{\mu}_r, \bar{\mu})$, where θ is the adsorbate coverage, and where $\bar{\mu}$ is the instantaneous potential during the increase back to saturation. It is often useful to calculate the FORC distribution,

$$\rho = -\frac{1}{2}\frac{\partial^2 \theta}{\partial \bar{\mu}_r \partial \bar{\mu}}, \tag{14.1}$$

which measures the sensitivity of the dynamics to the progress of reversal on the major loop. Note that to normalize the FORC distribution, the term $\frac{1}{2}\delta(\bar{\mu} - \bar{\mu}_r)\frac{\partial \theta(\bar{\mu}_r, \bar{\mu})}{\partial \bar{\mu}}|_{\bar{\mu} \to \bar{\mu}_r^+}$ must be added to (14.1) [6]. Here we consider the distribution only away from the line $\bar{\mu} = \bar{\mu}_r$. The additional term could be found from the major loop in Figs. 14.1a and 14.2a. The FORC distribution is usually displayed as a contour plot called a 'FORC diagram'.

14.2 Model

Kinetic Monte Carlo (KMC) simulations of a lattice-gas model, where a Monte Carlo (MC) step corresponds to an attempt to cross a local free-energy barrier, have been used to simulate the kinetics of electrochemical systems with first- [7, 8] or second-order [9, 10] phase transitions in two dimensions. Similar to the standard Ising Hamiltonian, a grand-canonical effective lattice-gas Hamiltonian [7, 10–13]

$$\mathcal{H} = -\sum_{i<j} \phi_{ij} c_i c_j - \bar{\mu} \sum_{i=1}^{N} c_i , \tag{14.2}$$

is used to describe the energy associated with a lattice-gas configuration. Here, $\sum_{i<j}$ is a sum over all pairs of sites, ϕ_{ij} are the lateral interaction energies between particles on the ith and jth lattice sites, $\bar{\mu}$ is the electrochemical potential, and $N = L^2$ is the total number of lattice sites for an $L \times L$ square lattice with periodic boundary conditions. The local occupation variable c_i is 1 if site i is occupied and 0 otherwise. In addition to adsorption/desorption steps, we include diffusion steps with a comparable free-energy barrier [10].

Independent of the diffusion, attractive nearest-neighbor interactions ($\phi_{ij} > 0$) produce a first-order phase transition equivalent to that which occurs below the Curie temperature in the ferromagnetic Ising model. In contrast, repulsive long-range interactions ($\phi_{ij} < 0$) and nearest-neighbor exclusion produce a second-order phase transition, equivalent to that in an antiferromagnetic Ising model.

14.3 First-order Phase Transition

Using an $L = 128$ lattice-gas model with attractive interactions, a family of FORCs were simulated, averaging over 10 KMC realizations for each reversal curve at room temperature. Using the notation of [10], the interaction

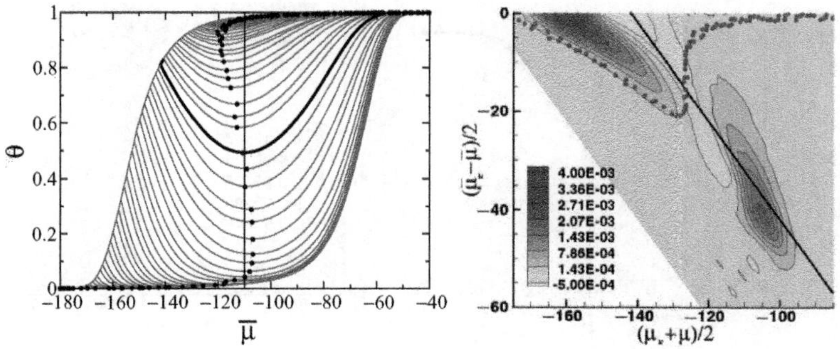

Fig. 14.1. (**a**) First-order reversal curves (FORCs) of a first-order phase transition. The *vertical line* shows the position of the coexistence value $\overline{\mu} = \overline{\mu}_c$. The minima of each FORC are also shown (*circles*). (**b**) FORC diagram generated from the family of FORCs shown in (**a**). The positions of the FORC minima are also shown (*circles*). The *straight line* corresponds to the FORC for which the minimum lies at the coexistence value $\overline{\mu} = \overline{\mu}_c$

strength (restricted to nearest-neighbor) was taken to be $\phi_{ij} = \phi_{nn} = 55\,\mathrm{meV}$. The barriers for adsorption/desorption and diffusion (nearest-neighbor only) were $\Delta_{a/d} = \Delta_{nn} = 150\,\mathrm{meV}$. The reversal electrochemical potentials $\overline{\mu}_r$ associated with the reversal curves were separated by 1 meV increments in the interval $[-200, 0]$, and the potential-sweep rate was constant at $|d\overline{\mu}/dt| = 0.3\,\mathrm{meV/MCSS}$. The FORCs are shown in Fig. 14.1a, with a vertical line indicating the position of the coexistence electrochemical potential, $\overline{\mu}_c$, and circles showing the positions of the minima.

In a simple Avrami's-law analysis, the FORC minima all lie at $\overline{\mu} = \overline{\mu}_c$ [5]. However, in the simulations the minima are displaced. For $\theta > 0.5$, the minima occur at $\overline{\mu} < \overline{\mu}_c$, precisely at the points where the tendency to phase order, which drives local regions of the system toward the nearby metastable state ($\theta \approx 1$), is momentarily balanced by the electrochemical potential, which drives the system toward the distant stable state ($\theta \approx 0$). For $\theta < 0.5$, the stable and metastable states are $\theta \approx 1$ and $\theta \approx 0$, respectively, and the same analysis explains the FORC minima occurring at $\overline{\mu} > \overline{\mu}_c$.

The net effect is a 'back-bending' of the curve connecting the minima, as seen in Fig. 14.1a. The definition in (14.1) implies that the FORC distribution ρ is negative in the vicinity of the back-bending as seen in Fig. 14.1b, where the FORC distribution is plotted against the variables $\overline{\mu}_b = (\overline{\mu}_r + \overline{\mu})/2$ and $\overline{\mu}_c = (\overline{\mu}_r - \overline{\mu})/2$. The negative values of ρ reflect a local divergence of the FORCs, which can be considered a dynamical instability, caused by the competition between the tendency to phase order and the effect of the potential. It is interesting to note that the curve connecting the minima of the FORCs resembles a van der Waals loop, but with an asymmetrical shape about the point ($\overline{\mu} = \overline{\mu}_c, \theta = 0.5$), and with a sweep-rate dependent shape (not shown).

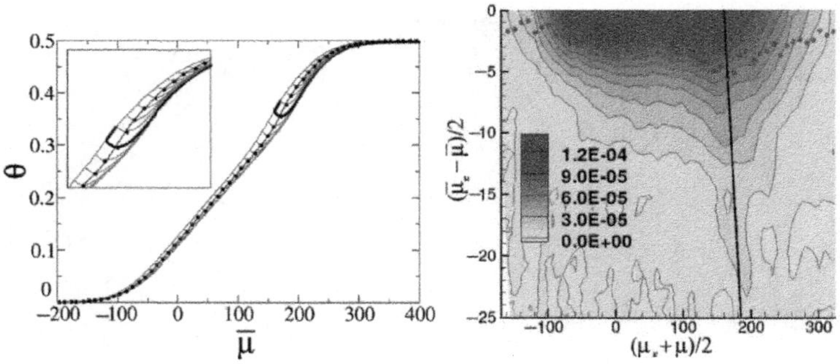

Fig. 14.2. (**a**) FORCs for a second-order phase transition. The *black middle line* shows the equilibrium curve. The *inset* is a magnification of the critical region. The minima of each FORC are also shown (*black circles*). (**b**) FORC diagram generated from the FORCs shown in (**a**). The positions of the FORC minima are also shown (*circles*). The *straight line* corresponds to the FORC for which the minimum lies closest to the critical coverage

We plan to explore the connection with the van der Waals loop in simulations of a mean-field model in future work.

14.4 Second-order Phase Transition

Using the same Hamiltonian, but with long-range repulsive interactions and nearest-neighbor exclusion, KMC simulations were used to produce the FORCs for a second-order phase transition. The reversal potentials $\overline{\mu}_r$ were separated by 10 meV increments in the interval $[-200, 400]$. As in [10], the repulsive $1/r^3$ interactions, with nearest-neighbor exclusion and $\phi_{nnn} = -21$ meV, are calculated with exact contributions for $r_{ij} \leq 3$, and using a mean-field approximation for $r_{ij} > 3$. The barriers for adsorption/desorption, and nearest- and next-nearest-neighbor diffusion, are $\Delta_{a/d} = 300$ meV, $\Delta_{nn} = 100$ meV, and $\Delta_{nnn} = 200$ meV. A second-order phase transition occurs between a disordered state at low coverage and an ordered state at high coverage [9, 10]. The FORCs and FORC diagram are shown in Fig. 14.2. Also indicated in Fig. 14.2a are the FORC minima and the equilibrium isotherm.

An important distinction here is that the FORC minima lie directly on the equilibrium isotherm. This is because such a system has one stable state for any given value of the potential, defined by the continuous equilibrium curve. The positive values of the FORC distribution in Fig. 14.2b are consistent with relaxation toward this equilibrium curve, at a rate which increases with the distance from equilibrium. It is interesting to note that, while it is difficult to see in Fig. 14.2a, the rate of approach to equilibrium decreases greatly along the first FORC that dips below the critical coverage $\theta_c = 0.36$ (shown in bold in Fig. 14.2a). This is clearly revealed in the FORC diagram in Fig. 14.2b, where it is reflected by the Florida-shaped "peninsula" centered around this FORC.

14.5 Comparison and Conclusions

Two observations can be made by comparing the FORCs and FORC diagrams for systems with first- and second-order phase transitions. First, the FORC minima in systems with a second-order phase transition correspond to the equilibrium behavior, while for systems with a first-order phase transition they represent van der Waals-like metastable and unstable regions. Thus, FORCs can be used to recover equilibrium behavior for systems with a second-order transition which need a long time to equilibrate. Second, due to the instability that exists in systems with a first-order phase transition, the corresponding FORC diagram contains negative regions which do not exist for systems with a second-order phase transition.

Acknowledgments

This research was supported by U.S. NSF Grant No. DMR-0240078, and by Florida State University through the School of Computational Science, the Center for Materials Research and Technology, and the National High Magnetic Field Laboratory.

References

1. I.D. Mayergoyz, IEEE Trans. Magn. **MAG 22**, 603 (1986).
2. C.R. Pike, A.P. Roberts, K.L. Verosub, J. Appl. Phys. **85**, 6660 (1999).
3. C. Enachescu, R. Tanasa, A. Stancu, F. Varret, J. Linares, E. Codjovi, Phys. Rev. B **72**, 054413 (2005).
4. M. Fecioru-Morariu, D. Ricinschi, P. Postolache, C.E. Ciomaga, A. Stancu, L. Mitoseriu, J. Optoelectron. Adv. Mater. **6**, 1059 (2004).
5. D.T. Robb, M.A. Novotny, P.A. Rikvold, J. Appl. Phys. **97**, 10E510 (2005).
6. C.R. Pike, Phys. Rev. B **68**, 104424 (2003).
7. S. Frank, D.E. Roberts, P.A. Rikvold, J. Chem. Phys. **122**, 064705 (2005).
8. S. Frank, P.A. Rikvold, Surf. Sci. (2006), in press.
9. S.J. Mitchell, S. Wang, P.A. Rikvold, Faraday Disc. **121**, 53 (2002).
10. I.A. Hamad, P.A. Rikvold, G. Brown, Surf. Sci. **572**, L355 (2004).
11. C.N. Yang, T.D. Lee, Phys. Rev. **87**, 410 (1952).
12. I.A. Hamad, T. Wandlowski, G. Brown, P.A. Rikvold, J. Electroanal. Chem. **554-555**, 211 (2003).
13. I.A. Hamad, S.J. Mitchell, T. Wandlowski, P.A. Rikvold, Electrochim. Acta **50**, 5518 (2005).

Part IV

Magnetic Systems

15

Vortex Fluctuation and Magnetic Friction

B.V. Costa[1], M. Rapini[1], R.A. Dias[2], and P.Z. Coura[1]

[1] Departamento de Física, ICEX, UFMG 30123-970 Belo Horizonte, MG, Brazil
[2] Departamento de Física, ICE, UFJF Juiz de Fora, MG, Brazil

Abstract. We use Monte Carlo and molecular dynamics simulation to study a magnetic tip-surface interaction. Our interest is to understand the mechanism of heat dissipation when the forces involved in the system are magnetic in essence. We consider a magnetic crystalline substrate composed of several layers interacting magnetically with a tip. The set is put thermally in equilibrium at temperature T by using a numerical Monte Carlo technique. By using that configuration we study its dynamical evolution by integrating numerically the equations of motion. Our results suggests that the heat dissipation in this system is closed related to the appearing of vortices in the sample.

The friction between two sliding surfaces is the result of many microscopic interactions between atoms and molecules of both surfaces, it dependents on the roughness, temperature and energy dissipation mechanisms. Therefore, to understand friction it is necessary to understand its microscopic mechanisms [1].

In general, in magnetic films which forms quasi 2 dimensional (2d) magnetic planar structures, the magnetization is confined to the plane due to shape anisotropy. An exception to that is the appearing of vortices in the system. To avoiding the high energetic cost of non-aligned moments, the vortex develops a three dimensional structure by turning out of the plane the magnetic moment components in the vortex core [2]. For data storage purposes, magnetic vortices are of high interest since its study provides fundamental insight in the mesoscopic magnetic structures of the system [3].

In this work we use a combined Monte Carlo-Molecular Dynamics (MC-MD) simulation to study the energy dissipation mechanism in a prototype model consisting of a reading head moving close to a magnetic disk surface. The model consists of a magnetic tip (The reading head) which moves close to a magnetic surface (The disk surface.). The tip is simulated as a cubic arrangement of magnetic dipoles and the surface is represented as a monolayer of magnetic dipoles distributed in a square lattice. We suppose that the dipole interactions are shielded, so that, we do not have to consider them as

Springer Proceedings in Physics, Volume 123
Computer Simulation Studies in Condensed-Matter Physics XIX
Eds.: D.P. Landau, S.P. Lewis and H.-B. Schüttler

long range interactions. The dipole can be represented by classical spin like variables $\boldsymbol{S} = S_x\hat{x} + S_y\hat{y} + S_z\hat{z}$. The total energy of this arrangement is a sum of exchange energy, anisotropy energy and the kinetic energy due to the relative movement between the tip and the surface as follows.

$$H = \sum_{\mathrm{i=1}}^{N_{\mathrm{h}}} \frac{\boldsymbol{p}^2_{(\mathrm{h-s})\mathrm{i}}}{2m_{(\mathrm{h})\mathrm{i}}} + U_{\mathrm{spin}} + U_{\mathrm{h-s}}, \tag{15.1}$$

where $U_{\mathrm{spin}} = U_{\mathrm{h}} + U_{\mathrm{s}}$.

$$U_{\mathrm{h}} = -\frac{J_{\mathrm{h}}}{2} \sum_{<\mathrm{i,j}>} (S^x_{\mathrm{hi}} \cdot S^x_{\mathrm{hj}} + S^y_{\mathrm{hi}} \cdot S^y_{\mathrm{hj}} + \lambda_h S^z_{\mathrm{hi}} \cdot S^z_{\mathrm{hj}}) - D_h \sum_{\mathrm{i=1}}^{N_{\mathrm{h}}} (S^z_{\mathrm{hi}})^2 \tag{15.2}$$

$$U_{\mathrm{s}} = -\frac{J_{\mathrm{s}}}{2} \sum_{<\mathrm{i,j}>} (S^x_{\mathrm{si}} \cdot S^x_{\mathrm{sj}} + S^y_{\mathrm{si}} \cdot S^y_{\mathrm{sj}} + \lambda_s S^z_{\mathrm{si}} \cdot S^z_{\mathrm{sj}}) - D_s \sum_{\mathrm{i=1}}^{N_{\mathrm{s}}} (S^z_{\mathrm{si}})^2 \tag{15.3}$$

and

$$U_{\mathrm{h-s}} = -\sum_{\mathrm{i,j}} J_{\mathrm{h-s}} \left(|\boldsymbol{r}_{\mathrm{hi}} - \boldsymbol{r}_{\mathrm{sj}}|\right) \left(\boldsymbol{S}_{\mathrm{hi}} \cdot \boldsymbol{S}_{\mathrm{sj}}\right) \tag{15.4}$$

with

$$J_{\mathrm{h-s}} \left(|\boldsymbol{r}_{\mathrm{hi}} - \boldsymbol{r}_{\mathrm{sj}}|\right) = J_0 \exp\{-\alpha \left(\boldsymbol{r}_{\mathrm{hi-sj}} - r_0\right)^2\} \tag{15.5}$$

In (15.1) the first term, $\boldsymbol{p}^2_{(\mathrm{h-s})\mathrm{i}}/2m_{(\mathrm{h})\mathrm{i}}$, stands for the relative kinetic energy: surface-reading head $(s-h)$. The second term, U_{spin}, accounts for the magnetic dipoles interactions: in the tip (U_{h}) and in the surface (U_{s}). The last term, $U_{\mathrm{h-s}}$, is the interaction energy between the tip and the surface. For the tip-surface interaction, we suppose that the coupling, $J_{\mathrm{h-s}}$, is ferromagnetic. By considering that $J_{\mathrm{h-s}}$ is a function of distance, will allow us to consider the effects of the relative tip-surface movement. The exchange anisotropy term λ, controls the kind of vortex which is more stable in the lattice. There is a critical value of the anisotropy, $\lambda_{\mathrm{c}} \approx 0.7\,\mathrm{J}$, such that for $\lambda < \lambda_{\mathrm{c}}$ the spins inside the vortex core minimizes the vortex energy by laying in an in-plane configuration. For $\lambda > \lambda_{\mathrm{c}}$ the configuration that minimizes the vortex energy is for the spins close to the center of the vortex to develop a large out-of-plane component [2].

The equations of motion [4] are solved by increasing forward in time the physical state of the system in small time steps of size $\delta t = 10^{-3}\,\mathrm{J}^{-1}$ by using Runge-Kutta's method of integration. With the head far from the surface, we start the time evolution of the system at $t = 0$ with $v = 0$. This part of the simulation serves as a guide to the rest of the simulation. Only thermal fluctuations of the vortex density can be seen. At $t = 200$ the tip is released with initial velocity v_0.

For the system with out-of-plane symmetry (Top line in Fig. 15.1.) we observe that for low temperature the vortex density augments when the tip passes over the surface. Initially the vortex density grows reaching quickly a

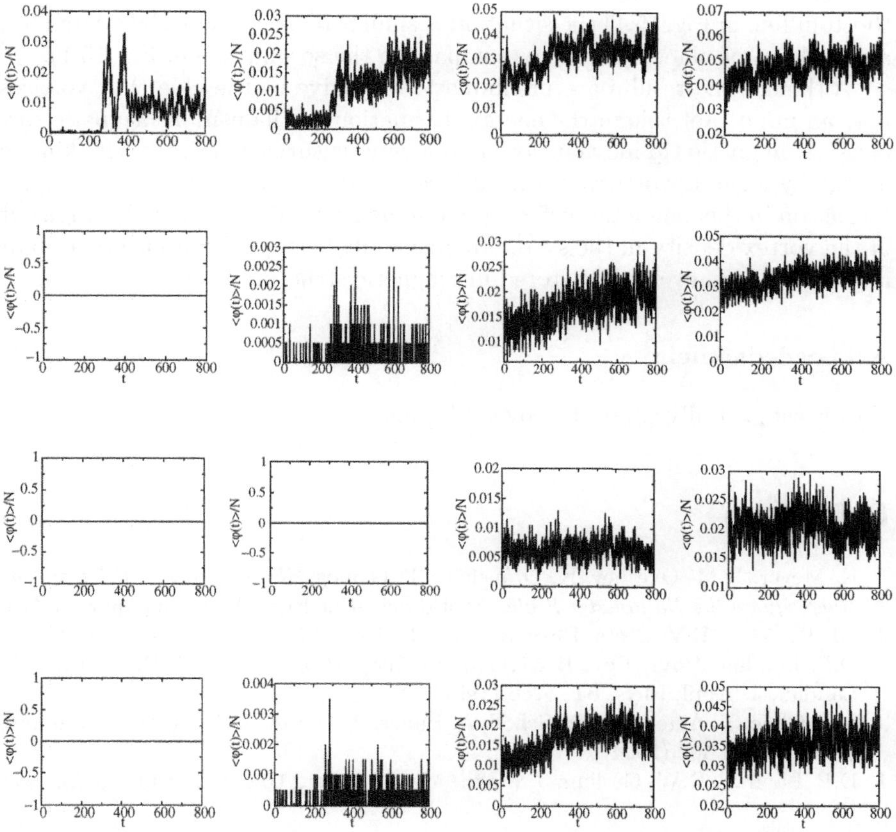

Fig. 15.1. Vortex Density as a function of time. From *top to bottom* and *left to right*: $D_{\rm h} = D_{\rm s} = 0.1J$ and $\lambda_{\rm s} = \lambda_{\rm s} = 1J$, $D_{\rm h} = D_{\rm s} = -0.1J$ and $\lambda_{\rm s} = \lambda_{\rm s} = 1J$, $\lambda_{\rm s} = \lambda_{\rm s} = 0.6J$ and $D_{\rm h} = D_{\rm s} = 0$ and $\lambda_{\rm s} = \lambda_{\rm s} = 0.9J$ and $D_{\rm h} = D_{\rm s} = 0$. Temperatures are $T = 0.1$, 0.2, 0.5, 0.6

constant average. For higher temperature the vortex density is almost insensitive to the tip indicating that the energy transfer becomes more difficult. At low temperature it is easier to excite the vortex mode since they have low creation energy due to the out-of-plane spin component. At higher temperature the system is already saturated and creating a new excitation demands more energy. For the in-plane symmetry (second line in Fig. 15.1) the situation is opposite. At low temperature there is no energy transfer to vortex modes. Creating a vortex demands much energy because the spin component of the vortex is almost whole in-plane. At higher temperature the system is soft, so that, it can absorb energy augmenting the vortex density. Eventually it reaches saturation at high enough temperature. For the case when the vortex has an in-plane-symmetry ($\lambda < \lambda_{\rm c}$), shown in the third line in Fig. 15.1, the average vortex density is constant even at higher temperatures. For ($\lambda > \lambda_{\rm c}$)

(bottom line in Fig. 15.1) the situation is similar to that where the system has a global out-of-plane symmetry (compare to the second line in Fig. 15.1).

Although the results are preliminary, they give us the clue that vortices play an important role in the energy dissipation mechanism in magnetic surfaces. A magnetic tip moving close to a magnetic surface, decreases its kinetic energy by forming vortices in the surface. On the point of view of storing information in the magnetic surface it is important to observe that the augment of the vortex density in the system increases its entropy. That effect can blur any information eventually stored in magnetic structures.

Acknowledgement

This work partially supported by CNPq and NSF.

References

1. E. Meyer, R.M. Overney, K. Dransfeld, T. Gyalog, *Nanoscience – Friction and Rheology on the Nanometer Scale.* (World Scientific Publishing, Singapore, 1998).
2. J.E.R. Costa, B.V. Costa, Phys. Rev. B **54**, 994 (1996); J.E.R. Costa, B.V. Costa, D.P. Landau, Phys. Rev. B **57**, 11510 (1998); B.V. Costa, J.E.R. Costa, D.P. Landau, J. Appl. Phys. **81**, 5746 (1997).
3. S.-B Choe, Y. Acremann, A. Scholl, A. Bauer, A. Doran, J. Stöhr, H.A. Padmore, Science **304**, 402 (2004).
4. D.P. Landau, R.W. Gerling, J. Magn. Magn. Mater. **104–107**, 843 (1992).

16

Simulational Study on the Linear Response for Huge Hamiltonians: Temperature Dependence of the ESR of a Nanomagnet

M. Machida[1], T. Iitaka[2], and S. Miyashita[3]

[1] Laboratory of Atomic and Solid State Physics, Cornell University, Ithaca NY 14853, USA
[2] Ebisuzaki Computational Astrophysics Laboratory, RIKEN (The Institute of Physical and Chemical Research), 2-1 Hirosawa, Wako, Saitama 351-0198, Japan
[3] Department of Physics, Graduate School of Science, The University of Tokyo, 7-3-1 Hongo, Bunkyo-ku, Tokyo 113-0033, Japan

Abstract. We construct a new numerical method of calculating the linear response in the Kubo formula for huge-Hamiltonian quantum systems. With the help of this method, the temperature dependence of the ESR intensity of the nanomagnet V_{15} is obtained. The calculated intensity show good agreement with the experimental data by Ajiro et al. [Physica B **329-333**, 1138 (2003)]. Our numerical method consists of two techniques: a random vector technique and the Chebyshev-polynomial expansion of exponential operators.

16.1 Introduction

Recently various nanomagnets or nanoscale molecular magnets have been synthesized as a nice toy material for experimental study on quantum phenomena such as tunneling and decoherence. In this paper, we will focus on a nanomagnet V_{15}. We numerically reproduce the experimental result of the electron spin resonance (ESR) of this nanomagnet by constructing a new numerical method of calculating the linear response in the Kubo formula.

The nanomagnet V_{15} is the complex of formula $K_6\left[V_{15}^{IV}As_6O_{42}\left(H_2O\right)\right]\cdot 8H_2O$. In the molecule, fifteen vanadium ions of spin 1/2 are placed almost on a sphere: three ions in the middle are sandwiched by two hexagons formed by the upper and lower twelve ions [1]. In order to study the properties of V_{15}, the ESR measurement has been done by Ajiro et al. [2] The interactions in the Hamiltonian of V_{15} are not fully determined. In particular, although different experimental [3–5] and theoretical [6, 7] works indicate the existence

Springer Proceedings in Physics, Volume 123
Computer Simulation Studies in Condensed-Matter Physics XIX
Eds.: D.P. Landau, S.P. Lewis and H.-B. Schüttler

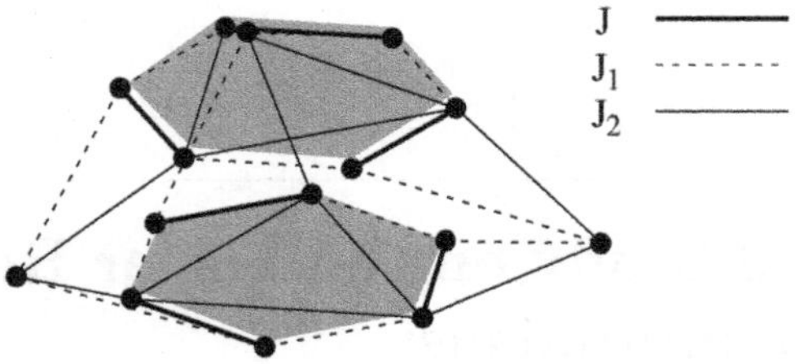

Fig. 16.1. The cluster of V_{15} showing the magnetic interactions J, J_1, and J_2

of the Dzyaloshinsky-Moriya interaction (DMI) [8–10] in V_{15}, the details of DMI are not yet fully understood.

16.2 ESR Intensity and Hamiltonian of V_{15}

Theoretically the ESR energy absorption $I(\omega;T)$ and intensity $I(T)$ are calculated by means of the Kubo formula. Here T is the temperature and ω is the frequency of the radiation magnetic field $\boldsymbol{H}_{\mathrm{R}}$ perpendicular to the static field $\boldsymbol{H}_{\mathrm{S}}$. $I(\omega;T)$ is given by the imaginary part of the dynamical susceptibility $\chi''(\omega;T)$:

$$I(\omega;T) = \frac{\omega H_{\mathrm{R}}^2}{2}\chi''(\omega;T). \tag{16.1}$$

The intensity is then given by $I(\omega;T)$:

$$I(T) = \int_0^\infty I(\omega;T)\,\mathrm{d}\omega. \tag{16.2}$$

We assume the following Hamiltonian for V_{15} [11,12,14].

$$\mathcal{H} = -\sum_{\langle i,j\rangle} J_{ij}\boldsymbol{S}_i\cdot\boldsymbol{S}_j + \sum_{\langle i,j\rangle}\boldsymbol{D}_{ij}\cdot[\boldsymbol{S}_i\times\boldsymbol{S}_j] - \sum_i \boldsymbol{H}_{\mathrm{S}}\cdot\boldsymbol{S}_i. \tag{16.3}$$

For J_{ij}, we have three different values J, J_1, and J_2 ($|J| > |J_2| > |J_1|$), as shown in Fig. 16.1. Here we take $J = -800\,\mathrm{K}$, $J_2 = -350\,\mathrm{K}$, and $J_1 = -225\,\mathrm{K}$ [13]. The second term on the right-hand side in (16.3) describes DMI. DM vectors are considered to exist on the two hexagons along the bonds with J. We take the reference DM vector $\boldsymbol{D}_{1,2}$ to be $D_{1,2}^x = D_{1,2}^y = D_{1,2}^z = 40\,\mathrm{K}$. The other DM vectors are determined from the D_3 symmetry of V_{15} [14]. We assume the static magnetic field is applied parallel to the c-axis of the molecule (z-axis) and put $\boldsymbol{H}_{\mathrm{S}} = (0,0,H_{\mathrm{S}})$. The magnitude of the static field H_{S} is 2 T (56.0 GHz).

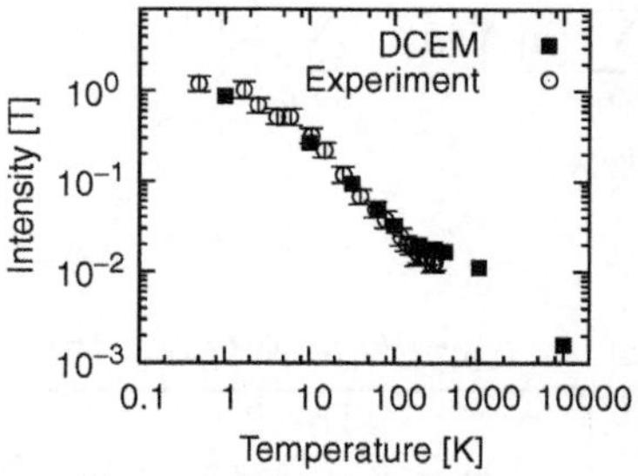

Fig. 16.2. The comparison between the numerical data calculated by DCEM (*solid squares*) and the experimental data by Ajiro et al. [2]

16.3 Double Chebyshev Expansion Method

In the Kubo formula, $\chi''(\omega;T)$ is calculated as follows.

$$\chi''(\omega;T) = \left(1 - \mathrm{e}^{-\beta\omega}\right) \mathrm{Re} \int_0^\infty \langle M^x M^x(t) \rangle \mathrm{e}^{-\mathrm{i}\omega t} \mathrm{d}t, \tag{16.4}$$

where $M^x = \sum_j S_j^x$. Since the size of Hamiltonians of nanomagnets is huge, it is difficult to calculate the spin-spin correlation function $\langle M^x M^x(t) \rangle$ in the above equation by direct diagonalization [15]. The dimension of the Hamiltonian for V_{15} is not very huge (= 32768), but it is still too large to diagonalize directly. Our method, which we call the double Chebyshev expansion method (DCEM) [14], makes it possible to calculate $\langle M^x M^x(t) \rangle$ of nanomagnets. This method is $O(N)$ (N is the dimension of the Hilbert space) for both memory size and computation time.

Figure 16.2 shows the temperature dependence of the intensity $I(T)$ of V_{15}. In the figure, solid squares are data obtained by DCEM whereas circles are data by the experiment [2]. We see that our result show good agreement with the experimental data. Our Hamiltonian (16.3) works very well. In what follows, we will explain our DCEM.

DCEM is an extension of the Boltzmann-weighted time-dependent method [18,19]. The difference between them is the time-evolution algorithm. DCEM consists of two techniques: a random vector technique and the Chebyshev-polynomial expansion of exponential operators.

First of all, we carry out the calculation of traces with the random vector technique. We prepare a random vector $|\Phi\rangle$ in the following manner [20,21]. For a given basis $|n\rangle$ of the Hilbert space, the random vector is given by

$$|\Phi\rangle = \sum_{n=1}^{N} |n\rangle \, \mathrm{e}^{\mathrm{i}\theta_n}, \tag{16.5}$$

where random numbers $\{\theta_n\}$ take values from $-\pi$ to π. Now we notice that a trace can be replaced by the ensemble averages ($[\cdot]_{\mathrm{av}}$) with respect to $|\Phi\rangle$: for any operator $\hat{X}$, $\mathrm{Tr}\hat{X}$ is calculated as

$$\left[\langle\Phi|\hat{X}|\Phi\rangle\right]_{\mathrm{av}} = \sum_{n}\langle n|\hat{X}|n\rangle + \sum_{m,n}\left[\mathrm{e}^{\mathrm{i}(\theta_m-\theta_n)} - \delta_{mn}\right]_{\mathrm{av}}\langle n|\hat{X}|m\rangle$$
$$\simeq \mathrm{Tr}\hat{X}. \tag{16.6}$$

Thus we have

$$\langle M^x M^x(t)\rangle = \frac{\mathrm{Tr}\mathrm{e}^{-\beta\mathcal{H}} M^x \mathrm{e}^{\mathrm{i}\mathcal{H}t} M^x \mathrm{e}^{-\mathrm{i}\mathcal{H}t}}{\mathrm{Tr}\mathrm{e}^{-\beta\mathcal{H}}}$$
$$= \frac{\left[\langle\Phi|\,\mathrm{e}^{-\beta\mathcal{H}/2} M^x M^x(t)\mathrm{e}^{-\beta\mathcal{H}/2}\,|\Phi\rangle\right]_{\mathrm{av}}}{\left[\langle\Phi|\,\mathrm{e}^{-\beta\mathcal{H}/2}\mathrm{e}^{-\beta\mathcal{H}/2}\,|\Phi\rangle\right]_{\mathrm{av}}}. \tag{16.7}$$

Next we expand the thermal operator $\mathrm{e}^{-\beta\mathcal{H}/2}$ and time-evolution operator $\mathrm{e}^{-\mathrm{i}\mathcal{H}t}$ with respect to the Chebyshev polynomials. To this end, we prepare a scaled Hamiltonian $\mathcal{H}_{\mathrm{sc}}$ ($= \mathcal{H}/\Delta\lambda$) so that the eigenvalues of $\mathcal{H}_{\mathrm{sc}}$ are confined between -1 to 1.

$$\mathrm{e}^{-\beta\mathcal{H}/2} = I_0\left(-\beta\Delta\lambda/2\right)T_0(\mathcal{H}_{\mathrm{sc}}) + 2\sum_{k} I_k\left(-\beta\Delta\lambda/2\right)T_k(\mathcal{H}_{\mathrm{sc}}),$$
$$\mathrm{e}^{-\mathrm{i}\tau\Delta\lambda\mathcal{H}_{\mathrm{sc}}} = J_0(\tau\Delta\lambda)T_0(\mathcal{H}_{\mathrm{sc}}) + 2\sum_{k}(-\mathrm{i})^k J_k(\tau\Delta\lambda)T_k(\mathcal{H}_{\mathrm{sc}}), \tag{16.8}$$

where $I_k(x)$ is the modified Bessel function, $J_k(x)$ is the Bessel function, and $T_k(\mathcal{H}_{\mathrm{sc}})$ is the Chebyshev polynomial, which satisfies $T_k(\mathcal{H}_{\mathrm{sc}}) = 2\mathcal{H}_{\mathrm{sc}}T_{k-1}(\mathcal{H}_{\mathrm{sc}}) - T_{k-2}(\mathcal{H}_{\mathrm{sc}})$, $T_0(\mathcal{H}_{\mathrm{sc}}) = 1$, and $T_1(\mathcal{H}_{\mathrm{sc}}) = \mathcal{H}_{\mathrm{sc}}$.

Since the time step τ in the time-evolution operator is not necessarily small, the time evolution by the Chebyshev-polynomial expansion works very well; in the ESR of V_{15}, the magnetic field H_{S} ($\sim 1\,\mathrm{K}$) is usually much smaller than the strongest coupling $|J|$ ($\sim 10^3\,\mathrm{K}$). Hence the frequency of precession of the spins is small. This means that we need to evolve state vectors for a long time but do not need fine resolution of the time step.

By these two techniques, we obtain the spin-spin correlation function without diagonalization.

Finally, $\chi''(\omega;T)$ is obtained by the Fourier transform of $\langle M^x M^x(t)\rangle$.

$$\chi''(\omega;T) = \left(1-\mathrm{e}^{-\beta\omega}\right)\mathrm{Re}\int_0^\infty \langle M^x M^x(t)\rangle \mathrm{e}^{-\mathrm{i}\omega t}\mathrm{d}t$$
$$= \left(1-\mathrm{e}^{-\beta\omega}\right)\mathrm{Re}\int_0^{T_{\max}} \langle M^x M^x(t)\rangle \mathrm{e}^{-\mathrm{i}\omega t}\mathrm{e}^{-\eta^2 t^2/2}\mathrm{d}t. \tag{16.9}$$

Here we introduced the Gaussian filter with variance $1/\eta^2$. This η determines the frequency resolution. The upper limit of the integral $T_{\max}$ satisfies $T_{\max} \sim 1/\eta$ in order to avoid the Gibbs oscillation. Also η should satisfy the conditions that $0 < \eta \ll 1$, $\eta \ll H_{\mathrm{S}}$, and $\beta\eta^2 \ll H_{\mathrm{S}}$.

Acknowledgement

This work is supported by a Grant-in-Aid from the Ministry of Education, Culture, Sports, Science and Technology, and also by NAREGI Nanoscience Project, Ministry of Education, Culture, Sports, Science and Technology, Japan. The simulations were partially carried out by using the computational facilities of the Super Computer Center of Institute for Solid State Physics, the University of Tokyo, and the Advanced Center for Computing and Communication, RIKEN (The Institute of Physical and Chemical Research).

References

1. D. Gatteschi, L. Pardi, A.L. Barra, A. Müller, J. Döring, Nature **354**, 463 (1991).
2. Y. Ajiro, Y. Inagaki, H. Itoh, T. Asano, Y. Narumi, K. Kindo, T. Sakon, H. Nojiri, M. Motokawa, A. Cornia, D. Gatteschi, A. Müller, B. Barbara, Physica B **329-333**, 1138 (2003).
3. I. Chiorescu, W. Wernsdorfer, A. Müller, H. Bögge, B. Barbara, Phys. Rev. Lett. **84**, 3454 (2000).
4. I. Chiorescu, W. Wernsdorfer, A. Müller, H. Bögge, B. Barbara, J. Mag. Mag. Mat. **221**, 103 (2000).
5. I. Chiorescu, W. Wernsdorfer, A. Müller, S. Miyashita, B. Barbara, Phys. Rev. B **67**, 020402 (2003).
6. K. Saito, S. Miyashita, J. Phys. Soc. Jpn. **70**, 3385 (2001).
7. S. Miyashita, N. Nagaosa, Prog. Theor. Phys. **106**, 533 (2001).
8. I. Dzyaloshinsky, J. Phys. Chem. Solids **4**, 241 (1958).
9. T. Moriya, Phys. Rev. Lett. **4**, 228 (1960).
10. T. Moriya, Phys. Rev. **120**, 91 (1960).
11. H. De Raedt, S. Miyashita, K. Michielsen, Phys. Stat. Sol. B **241**, 1180 (2004).
12. H. De Raedt, S. Miyashita, K. Michielsen, M. Machida, Phys. Rev. B **70**, 064401 (2004).
13. N.P. Konstantinidis, D. Coffey, Phys. Rev. B **66**, 174426 (2002).
14. M. Machida, T. Iitaka, S. Miyashita, J. Phys. Soc. Jpn. Suppl. **74**, 107 (2005).
15. S. Miyashita, T. Yoshino, A. Ogasahara, J. Phys. Soc. Jpn. **68**, 655 (1999).
16. T. Sakon, K. Koyama, M. Motokawa, Y. Ajiro, A. Müller, B. Barbara, Physica B **346-347**, 206 (2004).
17. M. Machida, T. Iitaka, S. Miyashita: in preparation.
18. T. Iitaka, S. Nomura, H. Hirayama, X. Zhao, Y. Aoyagi, T. Sugano, Phys. Rev. E **56**, 1222 (1997).
19. T. Iitaka, T. Ebisuzaki, Phys. Rev. lett. **90**, 047203 (2003).
20. A. Hams, H. De Raedt, Phys. Rev. E **62**, 4365 (2000).
21. T. Iitaka, T. Ebisuzaki, Phys. Rev. E **69**, 057701 (2004).

17

Attraction-limited Cluster–Cluster Aggregation of Ising Dipolar Particles

N. Yoshioka[1], I. Varga[2], F. Kun[2], S. Yukawa[1], and N. Ito[1]

[1] Department of Applied Physics, Graduate School of Engineering, The University of Tokyo, 7-3-1, Hongo, Bunkyo-ku, Tokyo, 113-8656, Japan
[2] Department of Theoretical Physics, University of Debrecen, P.O. Box 5, H-4010 Debrecen, Hungary

Abstract. The attraction-limited cluster–cluster aggregation of two-dimensional Ising dipolar particles with or without particle-size dispersity is studied. The fast decrease of the number of even-sized clusters for relatively smaller clusters is observed. Furthermore, it is suggested that, even in the dilute limit, the dynamic exponents are affected by the screening of the surrounding clusters on collision between two clusters.

17.1 Introduction

The structure formation in a monolayer of dipolar particles has been a subject of academic and practical interest for the past decades. Although there have been several studies of cluster–cluster aggregation (CCA) of dipolar particles [1–3], the comprehension of CCA of dipolar particles is far from complete. We recently examine the CCA of Ising dipolar particles (IDPs) [4], which is the particles confined to a plane and having Ising dipole moments, i.e., their dipole moments are constrained to be perpendicular to the plane [5,6]. The CCA of IDPs is a simple and realistic model of *attraction-limited* CCA (ALCA) of dipolar particles, or a limiting case in which thermal diffusion is ignored. The CCA of IDPs is also interesting as ALCA of *oppositely charged* particles, in connection with the heteroaggregation of oppositely charged colloids [7,8].

In this article, we discuss the dynamic properties of ALCA of IDPs. Especially, the cluster discrimination, i.e., the fast decrease of the number of even-sized clusters for relatively smaller clusters, and the dynamic exponents in dilute limit are considered.

Springer Proceedings in Physics, Volume 123
Computer Simulation Studies in Condensed-Matter Physics XIX
Eds.: D.P. Landau, S.P. Lewis and H.-B. Schüttler

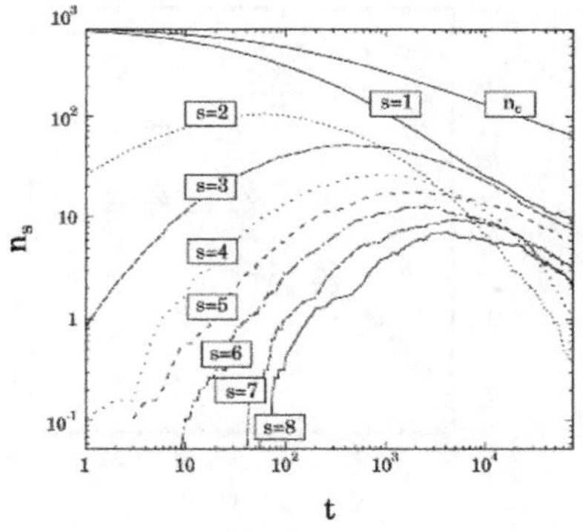

Fig. 17.1. Time evolution of the numbers of s-clusters for $\phi = 0.1$, $\sigma = 1.0$. The *thick solid line* denotes the total number of clusters

17.2 Methods

We treat an ensemble of $N = 750$ particles in a periodic square box. Each particle has a pointlike dipole moment in its center, and the dipole moment is fixed to be perpendicular to the plane of motion during the time evolution of the system. We consider only three forces which a particle $i \in \Pi_N$ experiences: the dipolar force, the Stokes force, and the Hertz contact force. The system is also supposed to be fully dissipative and deterministic. We perform molecular dynamics simulations for the system, and also perform the corresponding experiments. See [4] for more details.

17.3 Results

Here we summarize our results from the simulations and the experiments. For more details, see [4].

- For relatively smaller cluster sizes such as $s = 1, 2, 3$, and 4, the number of even-sized clusters decays faster than that of odd-sized ones (see Fig. 17.1). This behavior is known as *cluster discrimination* in CCA of oppositely charged colloids [7, 8].
- The average cluster size and the total number of clusters obey the power-law behavior: $S_{\text{av}}(t) \sim t^z$ and $N_{\text{c}}(t) \sim t^{-z'}$. The dynamic exponents z and z' as functions of the concentration ϕ are shown in Fig. 17.2.
- There exists a threshold concentration, $\phi_{\text{c}} = 0.05$, below which the dynamic scaling theory [9] can be applied. Above the threshold concentration, $z \neq z'$ and $z(\sigma = 1) \neq z(\sigma = 2.6)$ are observed, where σ is the

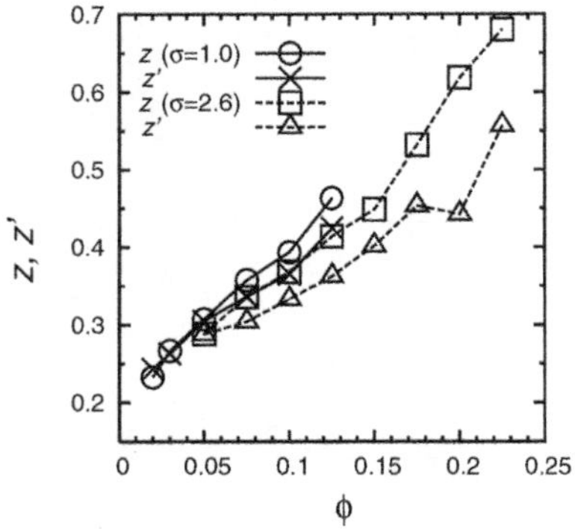

Fig. 17.2. Dependence of dynamic exponents z and z' on concentration ϕ for $\sigma = 1.0$ (*solid lines*) and $\sigma = 2.6$ (*dashed lines*) obtained from simulations. The error bars are omitted because the statistical errors of the exponents are small enough for the size of the error bars to fall in the order of the size of the symbols

particle-size dispersity, i.e., the ratio of the radii of larger IDPs to those of smaller IDPs.

17.4 Discussions

17.4.1 Cluster Discrimination

The cluster discrimination seems to contradict intuition; Imagine that the clusters are at long distances. In this situation, we can consider the effective potential for odd- and even-sized clusters such as $\sim r^{-3}$ and $\sim r^{-4}$. Therefore, we intuitively expect that attractions of odd-sized clusters is stronger and then decreases of the number of odd-sized clusters is faster.

The cluster discrimination is explained by the difference of *reactivities*: Since smaller clusters have chainlike structures, they can join other clusters only at chain ends. Odd-sized clusters, however, can form aggregates only with clusters that have *oppositely charged* particles, at least, at one of their ends because they have *equally charged* particles at both ends. In contrast, even-sized clusters have *differently charged* particles at both ends. Therefore, the *reactivities* of even-sized clusters are higher than those of odd-sized ones.

17.4.2 Dynamic Exponents in Dilute Limit

In Fig. 17.2, it appears that the dynamic exponents become close to 0.2 in the dilute limit. This value is much smaller than that of other CCAs, particularly that of ALCA of electrorheological (ER) suspensions in a strong electric field [1]. Here we consider the physical origin of this slow dynamics. First, the theoretical analysis by See and Doi [1] is briefly reviewed. Then we discuss whether their analysis can be applied to our systems.

Hierarchical-model-based Approach

They used a *hierarchical model* [10], that is, they assumed that sizes of all clusters were same, $s_k = 2^k$, for any k-th step, and only the two-body collision between nearest neighbor 2^k-clusters was considered at the k-th step.

Their system is constituted by chain-like aggregates, and dynamics of their system is nearly one-dimensional. Therefore, the orientation dependence of forces between the nearest neighbor 2^k-clusters are negligible. The potential energy between the two clusters is

$$\Phi_{\mathrm{SD}}(r) = -\sum_{n=1}^{s_k}\sum_{m=1}^{s_k} \frac{2\mu^2}{\left[(2n-1)R + (2m-1)R - 2^{k+1}R + r\right]^3}, \tag{17.1}$$

where r is distance between centers of the two clusters, and here it is assumed that all dipole moment and radii of particles are same, μ and R. If $s_k \gg 1$, then the summation in (17.1) can be transformed into the integral;

$$\Phi_{\mathrm{SD}}(r) \sim -\frac{2\mu^2}{r^3}\int_1^{\phi^{-1}} \mathrm{d}n \int_1^{\phi^{-1}} \mathrm{d}m \left[\frac{(2n-1)R}{r} + \frac{(2m-1)R}{r} - \frac{2^{k+1}R}{r} + 1\right]^{-3}. \tag{17.2}$$

Note that the supremum of the integral region in (17.2) is ϕ^{-1}; This is because the supremum is determined by the condition that we consider the time region during which the power-law behavior is observed. This condition implies that the average distance between clusters,

$$d = \left(\frac{L^2}{N_{\mathrm{c}}}\right)^{1/2} \sim \left(\frac{N\pi R^2/\phi}{N/s_k}\right)^{1/2} = \pi^{1/2} R \phi^{-1/2} s_k^{1/2},$$

is of the same order as the gyration radii of clusters, $R_{\mathrm{g}} \sim sR$, i.e., $s_k \sim \phi^{-1}$. This supremum is different from what See and Doi did [1]; They assumed that the supremum is infinity. However, the infinite supremum means that gelation ($d \ll R_{\mathrm{g}}$) regime is also considered.

Putting $x = (2n-1)R/r$ and $y = (2m-1)R/r$ in (17.2), and using $r > 2^{k+1}R$, we obtain

$$\Phi_{\mathrm{SD}}(r) \sim -\frac{\mu^2}{2R^2 r}\int_0^1 \mathrm{d}x \int_0^1 \mathrm{d}y\, (x+y+c)^{-3} = -\frac{C\mu^2}{2R^2 r},$$

where $0 < c < 1$ and C are constant. The equation of motion of the two clusters is

$$6\pi\eta s_k R\frac{\mathrm{d}r}{\mathrm{d}t} = -\frac{C\mu^2}{2R^2 r^2},$$

where η is the viscosity. Therefore, the time span t_k until which two clusters collide at the k-th step is

$$t_k = \frac{4\pi\eta R^3}{C\mu^2} s_k d^3 = \frac{4\pi^{5/2}\eta R^6}{C\mu^2}\phi^{-3/2} s_k^{5/2} \sim s_k^{5/2}.$$

Furthermore, the total elapsed time is

$$t \sim t_k + t_{k-1} + \cdots + t_1 \sim s_k^{5/2},$$

i.e., $s \sim t^{2/5}$. The dynamic exponent $z = 2/5$ was in good agreement with their simulation results [1].

Applicability for IDPs in Dilute Limit

In dilute limit, we can consider that all clusters of IDPs are straight chains. The potential energy between two straight IDP chains of the same size $s_k = 2^k$ and of the center-of-mass distance r is expressed as

$$\Phi_{\mathrm{IDP}}(r) = \mu^2 \sum_{n=1}^{s_k} \sum_{m=1}^{s_k} \frac{(-1)^{n+m}}{[d_{nm}(r)]^3}, \tag{17.3}$$

where $d_{nm}(r)$ is the distance between the n-th IDP of one chain and the m-th IDP of the other. If the summation in (17.3) is transformed into the integral such as (17.2), $\Phi_{\mathrm{IDP}}(r) \sim 0$. Therefore, we can not apply the approximation to $\Phi_{\mathrm{IDP}}(r)$ like See and Doi did.

Here we consider another approximation [4]. If $r \gg s_k(R_+ + R_-)$ and angular dependence of interaction is negligible, then

$$\Phi_{\mathrm{IDP}}(r) \sim -\frac{[\mu s_k(R_+ + R_-)]^2}{r^5}.$$

However, according to the See-Doi approach, we obtain $z = 2/5$. This value deviates from what we obtain, $z(\phi \to 0) = 0.2$. This means that, in the case of the ALCA of IDPs, the surrounding clusters and their screening effect are not negligible.

17.5 Summary

In summary, we have proposed the CCA of IDPs as a model of ALCA, and have studied its dynamic features. We have found that the *reactivities* of even-sized clusters are higher than those of odd-sized ones. In the dilute limit, the dynamic exponents become close to 0.2. This is considered because of the screening effect of the surrounding clusters.

Acknowledgment

F.K. is grateful to the Japan Society for the Promotion of Science for generous support during a research stay in Japan. This work was partly supported by JSPS Grants-in-Aid No. 14080204 and No. 14740229. F.K. and I.V. was supported by OTKA Grant No. T049209 and a Hungarian-Chinese Intergovernmental Project No. CHN-14/04.

References

1. H. See, M. Doi, J. Phys. Soc. Jpn. **60**, 2778 (1991).
2. M.C. Miguel, R. Pastor-Satorras, Phys. Rev. E **59**, 826 (1999).
3. J. Čerák, G. Helgesen, A.T. Skjeltorp, Phys. Rev. E **70**, 031504 (2004).
4. N. Yoshioka, I. Varga, F. Kun, S. Yukawa, N. Ito, Phys. Rev. E **72**, 061403 (2005).
5. I. Varga, F. Kun, K.F. Pál: Phys. Rev. E **69**, 030501(R) (2004).
6. I. Varga, H. Yamada, F. Kun, H.-G. Matuttis, N. Ito, Phys. Rev. E **71**, 051405 (2005).
7. A.M. Puertas, A. Fernández-Barbero, F.J. de las Nieves, Physica A **304**, 340 (2002).
8. J.M. López-López, A. Schmitt, J. Callejas-Fernández, R. Hidalgo-Álvarez, Phys. Rev. E **69**, 011404 (2004).
9. T. Vicsek, F. Family, Phys. Rev. Lett. **52**, 1669 (1984).
10. R. Botet, R. Jullien, M. Kolb, J. Phys. A **17**, L75 (1984).

Part V

Biological and Soft Condensed Matter

18

Simulational Study of the Multiple States in Hippocampal Slices

T. Shimada

Aihara Complexity Modelling Project, ERATO, JST
M204 Komaba Open Laboratory, The University of Tokyo,
4-6-1 Komaba, Meguro-ku, Tokyo, Japan

Abstract. Networks consist of simple conductance based model neurons coupled by synaptic connections are simulated to study the multiple activity states recently found in rat hippocampal slices. It is found that the neurons, each of which is chosen not to have multistable property, can show multistable activity in the presence of the connection. Different topologies of the synaptic connectivity (random net and the lattice) gives complementary explanations.

18.1 Introduction

Experimental studies have shown that neurons have various internal states such as so-called UP and DOWN states [1]. Recently Fujisawa et al. showed that the neurons in rat hippocampal slices possess a graded multiple states which are characterized by the UP/DOWN states like behavior [2]. The transition among these states can be induced by the repeated injection of current into a single cell. They further showed that those states do not appear when the synaptic connections are turned off by drug. Hence those states are very likely to be maintained by the activity of the network.

Since neural information processing is thought to be greatly affected by those states, to understand how those states emerge is important for brain science. However, present understanding remains in a conceptual stage: mean field picture with phenomenological spins. The aim of this study is to get some hints beyond the conceptual picture by simulating a neuronal network connected by excitatory and inhibitory synapses.

18.2 Model

The system consists of a conductance-based point neuron models and the phenomenological models for synapses.

Springer Proceedings in Physics, Volume 123
Computer Simulation Studies in Condensed-Matter Physics XIX
Eds.: D.P. Landau, S.P. Lewis and H.-B. Schüttler

18.2.1 Neuron Model

We use the simple neocortical model proposed by Wilson [3]. The equation of the motion for the membrane potential v ($\times 100$ [mV]) and the recovery variable r in [msec] unit are

$$\dot{v} = -(17.8+47.6v+33.8v^2)(v-0.5) - 26r(v+0.95) + I_{\rm syn} + I_{\rm in}, \quad (18.1)$$

$$\dot{r} = \frac{1}{\tau_r}\{(1.24 + 3.7v + 3.2v^2) - r\}, \quad (18.2)$$

where $I_{\rm syn}$, $I_{\rm in}$, and τ_r are the total synaptic current, other input currents, and the time constant, respectively. We apply nonzero input current $I_{\rm in}$ to mimic the current injection into the cell, and to mimic the effect of the muscarinic receptor agonist used in the experiment. The time constant τ_r is set to 4.2 [msec] for excitatory neurons and 1.0 [msec] for inhibitory neurons. Although the original model have additional two variables that represent a voltage-dependent Ca^2+ current and a Ca^{2+}-dependent slow hyperpolarizing current to mimic human neurons better, we here remove these two variables for simplicity. The topology of the two-dimensional phase plane is same as the Hindmarsh-Rose model [4]: nullclines of v(cubic) and r(quadratic) intersect and form stable node, saddle, and unstable spiral point.

18.2.2 Synaptic Efficacy

Neurons are connected by phenomenological model synapses [5,6]. Each model synapse has short-term plasticity which is expressed by the following two variables

$$\dot{R} = \frac{(1-R)}{D} - Rw\delta((t - t_{\rm input}), \quad (18.3)$$

$$\dot{w} = \frac{U-w}{F} + U(1-w)\delta((t - t_{\rm input}), \quad (18.4)$$

where R and w represent the depression variable and the facilitation variable. The parameters U, F, and D are chosen to 0.5, 1000, and 800 respectively for excitatory synapses. For inhibitory Synapses, we set 0.2, 20, and 700 respectively [7]. The synaptic efficacy is defined by the product of these variables.

18.2.3 Synaptic Conductances

Since the model neuron has no spatial structure, the total synaptic current is simply calculated as

$$I_{\rm syn} = -g_{\rm AMPA}(v-0) \quad (18.5)$$

$$-g_{\rm NMDA}\frac{\{(v+0.8)/0.6\}^2}{1+\{(v+0.8)/0.6\}^2}(v-0) \quad (18.6)$$

$$-g_{\rm GABA_A}(v+0.7) \quad (18.7)$$

$$-g_{\rm GABA_B}(v+0.9). \quad (18.8)$$

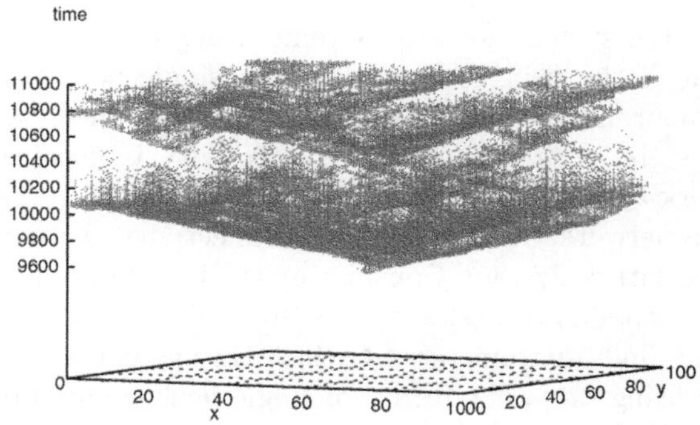

Fig. 18.1. A spatiotemporal raster plot of (100 × 100) lattice system. Traveling waves and transient burstings maintains the metastable active state

Each variable represents the total conductance of the channels of its type. Note that the reversal potential of the GABA-A channel (−70 [mV]), which is generally regarded as an 'inhibitory' channel, is higher than the resting potential (−74.8 [mV]).

The conductance decays as time in the absence of the synaptic input. When the neuron receives an input, it increases instantaneously by synaptic weight times the synaptic the efficacy:

$$\dot{g_\alpha} = -\frac{g_\alpha}{\tau_\alpha} + c_{ij} R_{ij} w_{ij} \delta(t - t_{\text{input}}). \tag{18.9}$$

Where c_{ij} is the synaptic weight between a presynaptic neuron i and a postsynaptic neuron j. The input time t_{input} is calculated as $t_i^{\text{fire}} + \text{axonal} - \text{delay}_{ij}$. Typical values of c_{ij}s is chosen to ~ 1 to get moderate ($1 \sim 10$ [mV]) amplitude of post synaptic potentials.

In the following simulation, the 4th order Runge-Kutta method is used for the neuronal variables v and r for time integration. For synaptic variables (g, r, w), we use 1st order Euler method. The time step is chosen to 0.1 [msec].

18.3 Results

From the size of the slice ($100 \sim 200$ [μm] ×300 [μm] ×$500 \sim 1000$ [μm]), the total number of the pyramidal neurons ($1,000 \sim 10,000$) and the ratio of the excitatory and inhibitory neurons ($\sim 80\%$ excitatory and $\sim 20\%$ inhibitory) can be estimated. However, the connectivity among them is unclear. Thus we first connect the neurons randomly. Current I_{in} is injected to a single neuron repeatedly according to the protocol of Fujisawa et al. In this case, the UP state like potential shift driven by the synaptic inputs can be reproduced in the plausible parameter region. However, what one can have in plausible parameter region is either the system which can not activated by the single cell bursting or the system goes to extreme firing state. The reason for this is rather trivial: if the firing or bursting of a single neuron can induce the firing of other neuron, the same process continues till the system goes to the explosion. Thus the clustering of the connectivity is necessary to have an operable network activity.

To simulate networks with correlated connectivity. we next consider the two dimensional lattice system. Considering the two dimensional geometry of the slice, it corresponds to assuming the spatially localized connectivity. In this case, we can find a parameters by which the system has metastable state with moderate firing rate. The firings of single neuron can induce the transition from the quiescence state to this metastable state. This state consists of transient traveling waves and localized firing (Fig. 18.1).

Recently the highly nonrandom and well clustered nature of the connectivity among neurons in real systems is clearly shown [8]. To include more realistic topology will be the next step. Other properties which may have to be included is the gap (diffusive) connections.

References

1. M. Striade, A. Nunez, F. Amzica, J. Neurosci. **13**, 3252 (1993).
2. S. Fujisawa, N. Matsuki, Y. Ikegaya, Cereb. Cort. in press.
3. H.R. Wilson, J. theor. Biol. **200**, 375 (1999).
4. J.L. Hindmarsh, R.M. Rose, Proc. R. Soc. Lond. B **221**, 87 (1984).
5. H. Markram, M. Tsodyks, Nature **382**, 807 (1996).
6. H. Markram, Y. Wang, M. Tsodyks, PNAS **95**, 5323 (1998).
7. E.M. Izhikivich, J.A. Gally, G.M. Edelman, Cereb. Cort. **14**, 933 (2004).
8. S. Sen *et al.*, PLoS Biology **3**, 507 (2005).

19

Dissipative Particle Dynamics of Self-Assembled Multi-Component Lipid Membranes

M. Laradji[1,3] and P.B. Sunil Kumar[2,3]

[1] Department of Physics, The University of Memphis, Memphis, TN 38152, USA
[2] Department of Physics, Indian Institute of Technology Madras
Chennai 600036, India
[3] MEMPHYS-Center for Biomembrane Physics, University of Southern Denmark
DK-5230, Denmark

Abstract. Self-assembled multicomponent lipid vesicles are studied via large scale dissipative particle dynamics simulations. We investigated the effect of volume fraction, line tension, surface tension, and transbilayer asymmetry in the lipid distribution on the dynamics and morphology of the membrane. We found that in the of symmetric transbilayer lipid distribution, the dynamics is rich characterized by coalescence of flat patches, budding and coalescence of caps. However, an asymmetric transbilayer lipid distribution sets a spontaneous curvature and lead to dramatically slow dynamics at intermediate values of the surface tension.

19.1 Introduction

Due to their strong hydrophobic interactions, amphiphilic molecules spontaneously aggregate in water to form a variety of self-assemblies [1]. Biomembranes, in particular, are naturally ocuring self-assemblies of a variety of lipid molecules and cholesterol [2]. Self-assembled biomembranes play two major roles: First, they serve as a topological separation between the inner and outer compartments of the cell or inner organelles (such as Golgi apartus, mitochondria, and the nucleus), and second, they act as a support of a complex protein based machinery which is important for various physiological functions, transmembrane transport, and the structural integrity of the cell [2]. Lipid membranes in the form of closed vesicles or supported bilayers are also synthesized and are used to understand biological membranes, and have a variety of biomedical and technological applications. In contrast to the Singer-Nicholson mosaic model [3] for lipid membranes, which assumes that biomembranes are homogeneous structures, many recent experiments demonstrated that biomembranes of eukaryotic cells exhibit both compositional and conformational organization. In particular, it was recently demonstrated that plasma

Springer Proceedings in Physics, Volume 123
Computer Simulation Studies in Condensed-Matter Physics XIX
Eds.: D.P. Landau, S.P. Lewis and H.-B. Schüttler

membranes of mammalian cells are laterally organized into small nanoscale domains, commonly referred to as lipid rafts, which are rich in sphingomyelin, which is a saturated lipid, and cholesterol. These domains are surrounded by a see of mainly non-saturated lipids. Lipid rafts are believed to be essential for physiological functions such as signaling, the recruitment of specific proteins and endocytosis [4]. Up to date, the very fundamental issues regarding the mechanisms leading to their formation and the stability of their finite size remain very unclear.

Lipid membranes at physiological conditions are very complex systems. Their structural complexity is further complicated by the presence of various types of vicinal molecules and their non-equilibrium activity. As such, there could be many mechanisms responsible for the formation of lipid rafts and their stability. Since biomembranes are multicomponent quasi-two-dimensional systems, we investigated a simple model for multicomponent lipid membranes using dissipative particle dynamics [5–7]. We will show that the dynamics of domain growth is inherently different from that of multicomponent Eucledean systems, and is affected by the interplay between composition, line tension between domains and the lateral tension on the membrane [5, 6]. In the case where the membrane is characterized by transbilayer asymmetry, as is the case of plasma membranes, we found that for intermediate lateral tensions, domain growth becomes exceedingly slow at late times leading to a microphase-separated domain structure [7]. The later result may explain the stability of lipid rafts in plasma membranes of mammalian cells.

19.2 Dissipative Particle Dynamics Approach for Multicomponent Lipid Membranes

The dynamics of phase separation in multicomponent lipid membranes was investigated in the past using a generalized time-dependent Ginzburg-Landau model on a non-Euclidean surface [8]. However, since only small deformations were allowed important processes such as budding and fission were not accounted for by this approach. A subsequent approach based on a dynamic triangulation model [9,10] predicted much richer dynamics than that obtained from the time-dependent Ginzburg-Landau model. The following comments should be made with respect to these two approaches: (1) the lipid membrane is not a self-assembled object, (2) they do account for an explicit solvent and are fully dissipative (no hydrodynamics), (3) they do not account for the constraint of volume enclosed by the vesicle, and (3) they conserve the topology of the membrane since they do not allow for fission or fusion events. In order to investigate the dynamics of phase separation of multicomponent vesicles, while accounting for these effects, we used the dissipative particle dynamics (DPD) approach.

The DPD approach, introduced by Hooggerbrugge and Koelman [11] and cast in its present form by Espanol [12] is a powerful numerical technique that has shown to be very useful in the understanding of both equilibrium and dynamics of complex fluids. What essentially differentiates the DPD method from molecular dynamics is that (1) in DPD, a number of atoms are coarse-grained to form a fluid element, thereafter called a dpd particle, (2) the dpd particles interact with each other via soft interactions, thereby allowing for longer integration times than in molecular dynamics, and (3) the use of pair-wise dissipative and random forces, in addition to the usual conservative forces. The dissipative and random forces, together, serve as a thermostat and their pairwise nature conserves local momentum, thereby leading to correct hydrodynamics. A dpd particle i experiences the following net force from other particles,

$$\boldsymbol{F}_{\rm i} = \sum_{\rm i \neq j} \left(\boldsymbol{F}_{\rm ij}^{\rm (C)} + \boldsymbol{F}_{\rm ij}^{\rm (D)} + \boldsymbol{F}_{\rm ij}^{\rm (R)} \right) , \tag{19.1}$$

where the conservative, dissipative and random forces are respectively given by

$$\boldsymbol{F}_{\rm ij}^{\rm (C)} = a_{\rm ij}\omega(r_{\rm ij})\hat{\boldsymbol{r}}_{\rm ij} , \tag{19.2}$$

$$\boldsymbol{F}_{\rm ij}^{\rm (D)} = \gamma_{\rm ij}\omega^2(r_{\rm ij})(\hat{\boldsymbol{r}}_{\rm ij} \cdot \boldsymbol{v}_{\rm ij})\hat{\boldsymbol{r}}_{\rm ij} , \tag{19.3}$$

$$\boldsymbol{F}_{\rm ij}^{\rm (R)} = \sigma_{\rm ij}(\Delta t)^{1/2}\omega(r_{\rm ij})\zeta_{\rm ij}\hat{\boldsymbol{r}}_{\rm ij} . \tag{19.4}$$

where $\hat{\boldsymbol{r}}_{\rm ij} = \boldsymbol{r}_{\rm j} - \boldsymbol{r}_{\rm i}$, $\hat{\boldsymbol{r}}_{\rm ij} = \boldsymbol{r}_{\rm ij}/r_{\rm ij}$, and $\boldsymbol{v}_{\rm ij} = \boldsymbol{v}_{\rm j} - \boldsymbol{v}_{\rm i}$. $\zeta_{\rm ij}$ is a symmetric random variable with zero mean and unit variance, uncorrelated for different pairs of particles and different times, i.e.,

$$\langle \zeta_{\rm ij}(t) \rangle = 0 , \tag{19.5}$$

$$\langle \zeta_{\rm ij}(t)\zeta_{\rm kl}(t') \rangle = (\delta_{\rm ik}\delta_{\rm jl} + \delta_{\rm il}\delta{\rm jk})\delta(t - t') , \tag{19.6}$$

with $i \neq j$ and $k \neq l$. The weight function, ω is given by

$$\omega(r) = \begin{cases} 1 - r/r_{\rm c} & \text{for } r \leq r_{\rm c}, \\ 0 & \text{for } r > r_{\rm c} \end{cases} \tag{19.7}$$

where $r_{\rm c}$ is a cutoff distance. The dissipative and random forces are related to each other through the fluctuation-dissipation theorem, leading to

$$\gamma_{\rm ij} = \sigma_{\rm ij}^2/2k_{\rm B}T . \tag{19.8}$$

The integrity of a lipid particle is ensured by an additional harmonic interaction given by

$$\boldsymbol{F}_{\rm i,i+1}^{\rm (S)} = -C\left(1 - \frac{r_{\rm i,i+1}}{b}\right)\hat{\boldsymbol{r}}_{\rm i,i+1} , \tag{19.9}$$

where C is some spring constant and b is some preferred bond length.

In our simulations, the system is composed of simple solvent particles (denoted as w), and two types of complex lipid particles (denoted as A and B lipids). A lipid particle is simulated as a linear and flexible amphiphilic chain with one hydrophilic dpd particle and a tail composed of three hydrophobic dpd particles. In our simulations, the interaction amplitude, a_{ij}, appearing in (19.2), between various types of dpd particles are selected as

$$a_{\rm ij} = \frac{\epsilon}{r_{\rm c}} \begin{pmatrix} & h_{\rm A} & t_{\rm A} & w & h_{\rm B} & t_{\rm B} \\ h_{\rm A} & 25 & 200 & 25 & a_{\rm AB} & 200 \\ t_{\rm A} & 200 & 25 & 200 & 200 & a_{\rm AB} \\ w & 25 & 200 & 25 & 25 & 200 \\ h_{\rm B} & a_{\rm AB} & 200 & 25 & 25 & 200 \\ t_{\rm B} & 200 & a_{\rm AB} & 200 & 200 & 25 \end{pmatrix} . \tag{19.10}$$

where energy and length scales are set by ϵ and $r_{\rm c}$, respectively. In (19.9), $C = 100\epsilon$ and $r_{\rm c} = 0.45r_{\rm c}$. We furthermore used $\sigma = 3.0(\epsilon^3 m/r_{\rm c}^2)^{1/4}$, a fluid number density $r_{\rm c} = 3.0r_{\rm c}^{-3}$ and the simulations were performed at $k_{\rm B}T = \epsilon$. The velocity-Verlet algorithm was used with an integration time step $\delta t = 0.05\tau$ with the time scale $\tau = (mr_{\rm c}^2/\epsilon)^{1/2}$. Our simulations are performed in cubic boxes of size $(80\times80\times80)r_{\rm c}^3$ corresponding to $1,536,000$ dpd particles, and we the total number of lipid particles is fixed at $16,000$.

Since our vesicles are relatively large, their formation from the self assembly of an initial homogeneous lipid solutions takes a large CPU time. In order to save computer time, almost closed spherical vesicles composed of A-lipids only are prepared. The vesicle with a pore is then let to equilibrate. This allows the lipid density in the inner and outer leaflets to equilibrate through the edge of the pore. Eventually the pore closes within 50,000 DPD steps. Prepared in this manner, the vesicle with our parameters contains about 138,500 solvent particles. The area-to-volume ratio is then controlled by transferring specific numbers of solvent particles from the core to the outer region of the vesicle. We define an area-to-volume ratio parameter $\nu = (N_{\rm head} + N_{\rm tail})/N_w$, where $N_{\rm head}$ and $N_{\rm tail}$ correspond to the numbers of lipid-head and lipid-tail particles, and $N_{\rm w}$ corresponds to the number of water particles.

We will present results for closed membranes with symmetric and asymmetric transbilayer lipid distributions. We furthermore investigated the effect of membrane tension, set by the area-to-volume constraint, and line tension set by the interaction between unlike lipid, $a_{\rm AB}$.

19.3 Phase Separation Dynamics in Two-Phase Closed Vesicles

In this section, we will first present results for the case of closed vesicles with symmetric transbilayer lipid distributions. In Fig. 19.1, snapshots of closed vesicles with a volume fraction of B-lipids, $\phi_{\rm B} = 0.3$, and is the same in both the inner and outer leaflets of the vesicle. This figure shows that domain growth is greatly affected by the the area-to-volume ratio. In the case of low excess area ($\nu = 0.462$) domains remains flat (i.e., they have the same curvature as the vesicle) throughout the phase separation process. In contrast, in the case of high excess area ($\nu = 0.567$) domains curve at intermediate times due to the presence of excess area. This is followed by the budding and vesiculation of some of the domains. Note that as the result of budding and vesiculation, the vesicle looses much of its excess area and becomes almost spherical.

In Fig. 19.2, the net length of the interface is shown for cases of high and low excess area of systems shown in Fig. 19.1. This figure demonstrates that during intermediate times, i.e., $t < 400\tau$, the excess area has no effect of domain growth. During this regime, domains have the same mean curvature as that of the surrounding majority component (see Fig. 19.1), and $L(t) \sim t^{-0.3}$. At later times, however, i.e. when $t > 400\tau$ the dynamics in the two systems become profoundly different. In the system with excess area, the interfacial length decreases rapidly due to budding and vesiculation of some of the B-domains. Once the vesicle loses much of its excess area, further budding of domains become prohibitive and the net interfacial length crosses over to a slower domain growth, characterized by $L(t) \sim t^{-0.45}$. The system with low excess area, domain growth remains the same as at earlier times, i.e., $L(t) \sim t^{-1/3}$.

In order to understand the origin of the power law, $L(t) \sim t^{-1/3}$, consider a vesicle composed of $N(t)$ B-domains with an average radius $R(t)$. The net interfacial length is therefore given by

$$L(t) = 2\pi R(t) N(t). \tag{19.11}$$

On the other hand, the net area occupied by B-domains, $\mathcal{A}_{\mathrm{B}}$, is given by

$$\mathcal{A}_{\mathrm{B}} = \pi N(t) R^2(t). \tag{19.12}$$

Therefore the net interfacial length and the number of domains are related to the average size of domains as

$$L(t) \sim \mathcal{A}_{\mathrm{B}}/R(t) \tag{19.13}$$

$$N(t) \sim \mathcal{A}_{\mathrm{B}}/R^2(t). \tag{19.14}$$

The net interfacial length versus time for systems shown in Fig. 19.1, is shown in Fig. 19.2. We note, again, that at relatively early times, domain growth in the two systems is consistent with the growth law, $L(t) \sim t^{-\alpha}$, with the growth exponent, $\alpha \approx 0.3$, very close to $\alpha = 1/3$. A growth exponent, $\alpha = 1/3$ is usually an indication that domain growth is the result of the evaporation-condensation mechanism of single lipids or small lipid clusters (Ostwald ripening theory due to Lifshitz and Slyozov) [13]. However, in our case, through careful monitoring, we found very few cases of single B-lipids diffusing in the A-lipid regions. Instead, domain growth is due to their Brownian motion in the majority component and their collision leading to coalescence. It is shown [5, 6] that the Brownian motion of circular lipid domains in a three-dimensional fluid, and coalescence yield to a growth law $R(t) \sim t^{1/3}$ and a

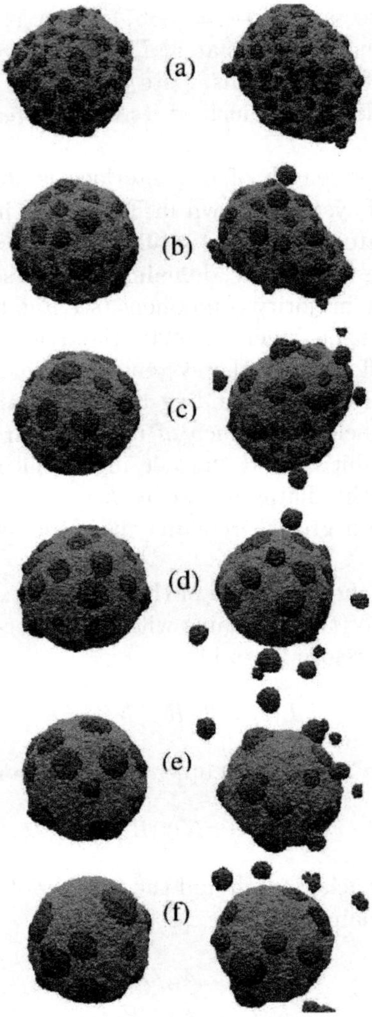

Fig. 19.1. Snapshots of phase-separating vesicles with the minority volume fraction, $\phi_B = 0.3$. Left and right columns corresponds to low excess area ($\nu = 0.567$) and high excess area ($\nu = 0.462$), respectively. The times correspond to (**a**) $t = 100\tau$, (**b**) 500τ, (**c**) 1000τ, (**d**) 2000τ, (**e**) 3000τ, and (**f**) 4000τ

number of clusters that decay as $N(t) \sim t^{-2/3}$ in agreement with our numerical findings. When domains are capped, using similar arguments, we showed that while $N(t) \sim t^{-2/3}$, $L(t) \sim t^{-4/9}$, again in very good agreement with our results for vesicles with excess area at late times.

In Fig. 19.3, the square of the fluctuations in the radius of vesicles with and without excess area are shown versus time. Note that when the excess area is very small, the fluctuations in the radius is very small throughout

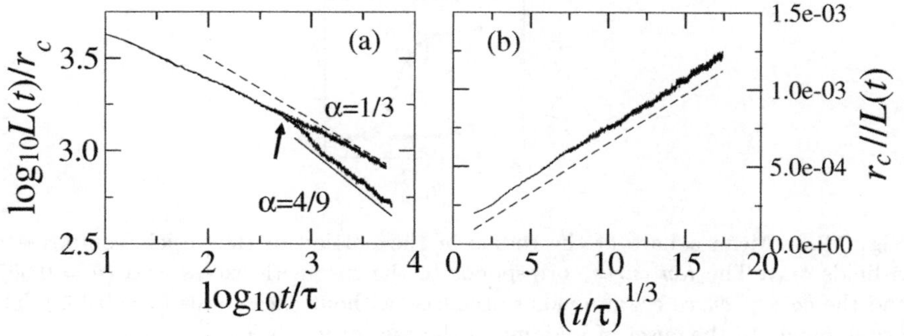

Fig. 19.2. (**a**) Net inerfacial length as a function of time in the case of $\phi_B = 0.3$, $\nu = 0.462$ (*top curve*) and $\nu = 0.567$ (*bottom curve*). The dashed and solid lines have slopes of $-1/3$ and $-4/9$, respectively. In graph (**b**), the inverse of the inerfacial length is shown versus $t^{1/3}$ for the case of $\nu = 0.462$. Notice the linear dependence of $1/L(t)$ versus $t^{1/3}$

the dynamics, and indication that the vesicle remains spherical throughout the dynamics, and the domains remain flat (i.e., have a curvature that is equal to that of the vesicle) through the phase separation process. In contrast the fluctuations are large during early times in the case of vesicles with excess area, an indication of a floppy vesicle. At later times, however, the fluctuations decrease. This is due to budding and fission of the domains from the parent vesicles, as indicated in Fig. 19.4. As fission proceeds, the parent vesicle loses much of it excess area, and the vesicle steadily acquires an almost spherical shape.

To understand the onset of this regime, we consider a B-lipid domain, of area a, a mean curvature c and an interfacial length l. We assume that the domain is embedded in A-lipid surrounding. The excess energy due to this domain, is given by

$$\Delta\mathcal{E} = 2\kappa a c^2 + \lambda l \,, \tag{19.15}$$

where κ and λ are the bending modulus and line tension, respectively. The perimeter of the domain, l is related to a and c through

$$l = 2\pi\left(\frac{a}{\pi}\right)^{1/2}\left(1 - \frac{ac^2}{4\pi}\right)^{1/2} . \tag{19.16}$$

The minimization of the excess energy, (19.15), yields a mean curvature of the domain, $c = 0$, i.e., a flat domain when the area of the domain, $a < a_0 = 4\pi(\kappa/\lambda)^2$. The onset of domain capping is $R_0 = 2\kappa/\lambda$. However, this domain capping necessitates a certain amount of excess area, i.e., a membrane with low surface tension. The estimated value for the crossover time as calculated from the equation above is found to agree very well with our simulation results.

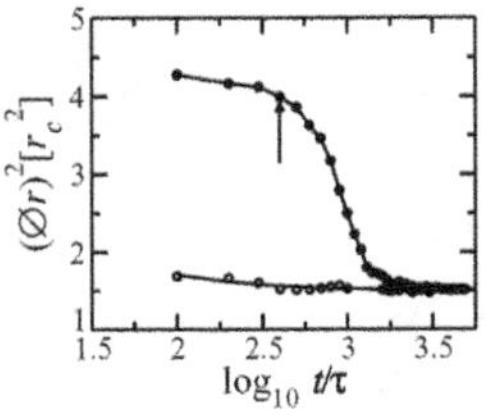

Fig. 19.3. Mean square of the radius of the portion of the vesicle composed of A-lipids only. The *top curve* corresponds to the case with excess area ($\nu = 0.567$) and the *bottom curve* corresponds to the case without excess area ($\nu = 0.462$). The *arrow* points to the onset of budding in the case of $\nu = 0.567$

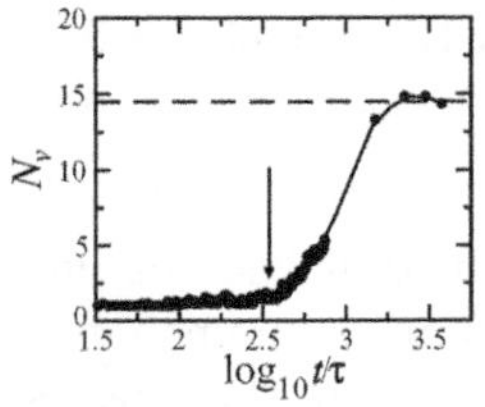

Fig. 19.4. The number of fissioned vesicles of the B-component versus time for the case with excess area ($\nu = 567$)

19.4 Effect of Line Tension on the Phase Separation

As discussed above, the interplay between line tension and bending modulus lead to different capping onset, and therefore different dynamics. In our simulations, the line tension, λ is changed by changing the interaction between A- and B-lipids while keeping the AA- and BB-interaction constant. We therefore carried simulations at $a_{\mathrm{AB}} = 50$, 68, and $100\epsilon/r_{\mathrm{c}}$. In Fig. 19.5, time sequences for the case of $\phi_{\mathrm{B}} = 0.3$ and an excess area corresponding to $\nu = 0.567$ are shown for these three line tensions. This figure clearly shows that line tension has a dramatic effect. We notice that while the configurations look similar at early times, we note that the higher is line tension, the earlier is domain capping, in agreement with the arguments presented in Sect. 19.4. We furthermore note from this figure that bud fission ocures only in the highest tension case (corresponding to $a_{\mathrm{AB}} = 100\epsilon/r_{\mathrm{c}}$). This is due to the fact that fission of a bud, which is a topological transformation, is accompanied by rearrangements of lipid molecules around the interface between the bud and the majority component. This is achieved if the excess energy available due to the interface ($\sim \lambda l$ where l is the length of the bud neck) is higher than energy barrier needed for to the lipid rearrangement. Therefore, as line tension is decreased, the energy barrier is too high to be overcome, and the domains

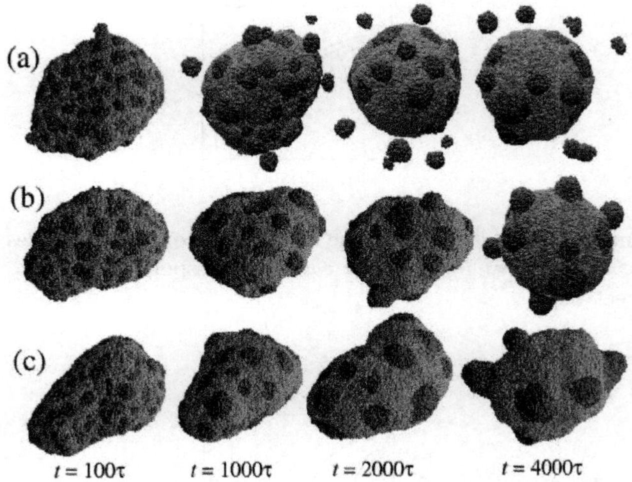

Fig. 19.5. Sequence of snapshots of vesicles with $\phi = 0.3$ and $\nu = 0.567$. (**a**) corresponds to $a_{\rm AB} = 100\epsilon/r_{\rm c}$, (**b**) corresponds to $a_{\rm AB} = 68\epsilon/r_{\rm c}$, and (**c**) corresponds to $a_{\rm AB} = 50\epsilon/r_{\rm c}$

remain attached to the vesicle as is the case for $a_{\rm AB} = 68$ and $50\epsilon/r_{\rm c}$ at late times. The net interfacial length for these three systems is shown versus time in Fig. 19.6, where we clearly note that while the dynamics is accompanied by a intermediate regime of bud fission in in the case $a_{\rm AB} = 100\epsilon/r_{\rm c}$, only two dynamical regimes, characterized by $\alpha = 1/3$ and $4/9$ are observed in the case $a_{\rm AB} = 50$ and $68\epsilon/r_{\rm c}$. We also notice in Fig. 19.6, that the crossover regime between these two regimes is delayed as the line tension is increased. This is due to the fact that the onset of domain capping occurs when $R \sim \kappa/\lambda$. Since before their capping domains grow as $t^{1/3}$, then the onset of capping occurs at $t_{\rm cr} \sim (\kappa/\lambda)^{1/3}$, and therefore is delayed as the line tension decreases, in qualitative agreement with Fig. 19.5. Note that as a result of a delay in capping as the line tension is decreased, the rate at which the vesicle's shape changes to spherical is also slowed down, as demonstrated by Fig. 19.7.

19.5 Effect of Asymmetry in the Composition of the Two Leaflets

Plasma membranes of mammalian cells are generically characterized by a marked transbilayer asymmetry in the lipid distribution. Indeed, the cytoplasmic leaflet contains mainly phosphatidylserine and phosphatidylethanolamine. The outer leaflet, however, contains mainly sphingomyelin and phosphatidyl-

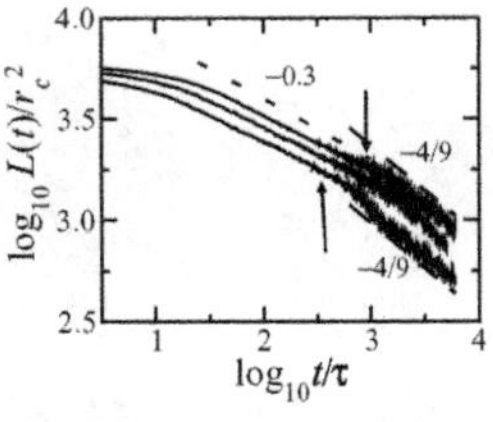

Fig. 19.6. Interfacial length versus time for the case of closed vesicles with excess area ($\nu = 0.567$). *Curves from top to bottom* correspond to $a_{\mathrm{AB}} = 50$, 68, and $100\epsilon/r_{\mathrm{c}}^2$, respectively

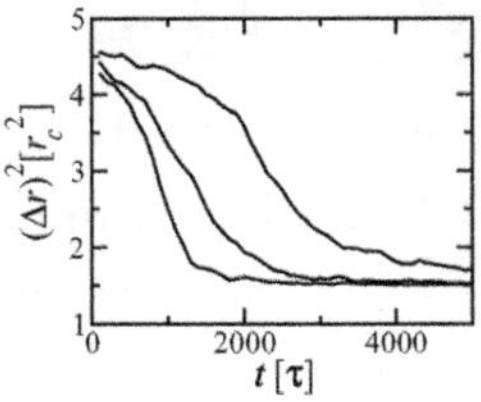

Fig. 19.7. Mean square of the radius of the portion of the vesicle composed of A-lipids only, for vesicles with $\nu = 0.567$. *Curves from top to bottom* correspond to $a_{\mathrm{AB}} = 50$, 68, and $100\epsilon/r_{\mathrm{c}}$, respectively

choline [2,14]. This asymmetry, believed to be maintained by the cell through many active processes, plays an important role to the lateral organization. In particular, recently discovered nanoscale domains, referred to as rafts have been indirectly observed in the outer leaflet of the plasma membrane of mammalian cells, and are composed mainly of sphingomyelin (a saturated lipid) and cholesterol [15,16]. The affinity between the saturated sphingomyelin and cholesterol is due to the conformational compatibility between the rigid backbone of cholesterol and the flexible acyl chains of sphingomyelin. Lipid rafts have been associated with physiological functions such as the recruitment of signaling proteins and endocytosis [4,15].

Synthesized lipid vesicles composed of a saturated lipid, an unsaturated lipid and cholesterol exhibit phase coexistence between a phase rich in the unsaturated lipid on one hand, and a phase rich in the saturated lipid and cholesterol, on the other hand [17–22]. The phase rich in the saturated lipid and cholesterol is reminiscent of lipid rafts. A question that arises is: Why are lipid rafts in the plasma membrane very small compared to lipid domains in synthesized vesicles? Although it is possible that the finite size of lipid rafts

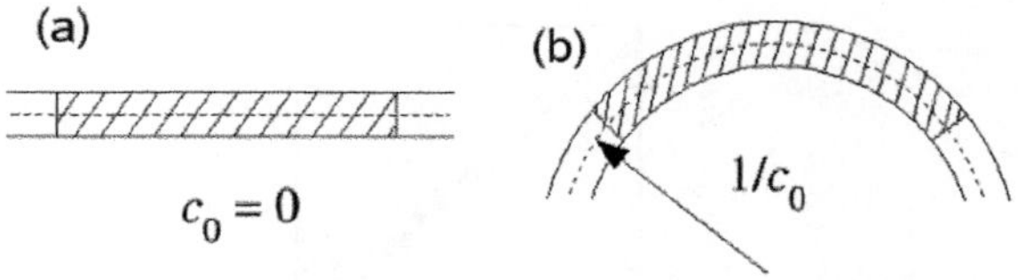

Fig. 19.8. Cross-section views of domain structures that minimize the interfacial (*line*) energy. (**a**) corresponds to the case where the lipid composition in the two leaflets is symmetric, and (**b**) corresponds to the case where the lipid composition between the two leaflets is asymmetric

Fig. 19.9. Sequence of snapshots of a system with $(\phi_{\rm out}, \phi_{\rm in}) = (0.3, 0.3)$ (**a**), and $(\phi_{\rm out}, phi_{\rm in}) = (0.4, 0.3)$ (**b**). Both systems have same overall lipid composition ($\phi = 0.3$), area-to-volume ratio parameter $\nu = 0.562$, and $a_{\rm AB} = 50\epsilon/r_{\rm c}$

is maintained by active processes, we will show that an asymmetry in the transbilayer lipid composition can also lead to curved finite size domains that could be reminiscent to lipid rafts.

In order to see how an asymmetry can lead to finite size domains, we must recall that several recent experiments on giant unilamellar vesicles demonstrated that segregated domains in the two leaflets of the membrane exhibit strong registry [20–23]. That is the lipids in the two leaflet are locally of same species. When a transbilayer asymmetry is present, the domains on the outer and inner leaflets must have different areas. Consequently, due to the fact that they also have to be in registry, these domains must be curved in order to minimize their line (interfacial) energy, as shown schematically in Fig. 19.8.

In Fig. 19.9, snapshots of a system with system with symmetric lipid distribution, $(\phi_{\rm out}, \phi_{\rm in}) = (0.3, 0.3)$ are shown against a system with same overall lipid composition but with $(\phi_{\rm out}, \phi_{\rm in}) = (0.4, 0.2)$. Both systems are at $\nu = 0.567$ and $a_{\rm AB} = 50\epsilon/r_{\rm c}$. We note from this figure that an asymmetry in the lipid distribution has a dramatic effect on the morphology of the vesicle.

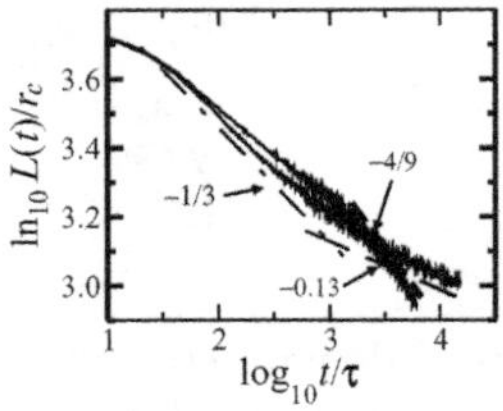

Fig. 19.10. Net interfacial length versus time. The *curve* with slope $-4/9$ at late times corresponds to $(\phi, out, \phi_{\rm in}) = (0.3, 0.3)$, and the *curve* with slope -0.13 at late times corresponds to $(\phi_{\rm out}, \phi_{\rm in}) = (0.4, 0.2)$

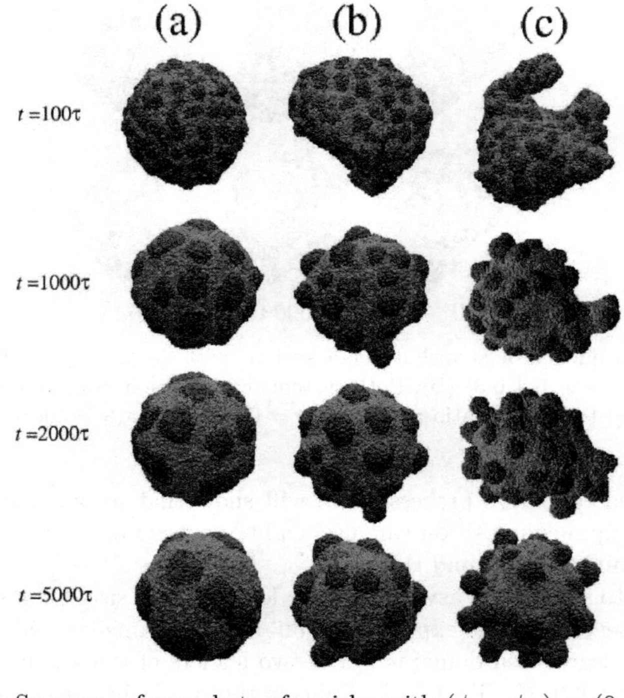

Fig. 19.11. Sequence of snapshots of vesicles with $(\phi_{\rm out}, \phi_{\rm in}) = (0.4, 0.2)$. **(a–c)** correspond to $\nu = 0.462$, $\nu = 0.567$, and $\nu = 0.882$, respectively. In all three systems, $a_{\rm AB} = 50\epsilon/r_{\rm c}$

In particular, we note that the onset of domain curvature occurs much earlier in time in the system with asymmetric transbilayer lipid distribution. This is due to the fact that domain capping in the case of transbilayer asymmetry is due to a spontaneous curvature generated by asymmetry in composition [7], while in the case of symmetric transbilayer composition, domain capping re-

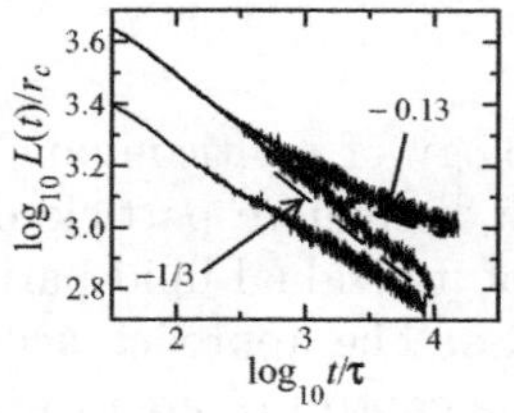

Fig. 19.12. Interfacial length versus time. *Curves from top to bottom* correspond to $\nu = 0.462$, $\nu = 0.567$ and $\nu = 0.882$, respectively. The data line for the case of $\nu = 0.462$ (*lowest curve*) has been shifted downward for clarity

sults from the interplay between the bending modulus and line tension [6]. The corresponding interfacial length versus time is shown in Fig. 19.10, where it is clear that the dynamics in the system with asymmetric transbilayer lipid distribution is dramatically slowed down at late times. This slowing down is an indication that the system undergoes a microphase separation instead of full phase separation as is the case of symmetric transbilayer lipid distribution.

In order to investigate the interplay between spontaneous curvature and surface tension, we performed simulations at three values of the area-to-volume ration, corresponding to $\nu = 0.462, 0.567$, and 0.882, and for $(\phi_{\text{out}}, \phi_{\text{in}}) = (0.4, 0.2)$. Snapshot sequences corresponding to these three systems are shown in Fig. 19.11. From these figures, we first notice that the vesicles acquire a spherical shape since early times, in contrast to the case where the compositions of the two leaflets are symmetric. In the case of medium and high excess area, the domains curve away from vesicle earlier that in the case with symmetric compositions, an indication that capping is now dominated by the spontaneous curvature, instead of the competition bending the bending modulus and the line tension. In the case of low excess area ($\nu = 0.462$), the domains curvature is almost equal to that of the vesicle, due to the lack of excess area.

The net interfacial length versus time for the three values of excess area is shown in Fig. 19.12. We note from this figure that while domain growth is relatively fast in the cases of low excess area ($\nu = 0.462$) and high excess area ($\nu = 0.882$), with a growth law, $L(t) \sim t^{-1/3}$, domain growth is much slower and non-algebraic in the case of intermediate excess area ($\nu = 0.567$). The slow dynamics in the case of intermediate tension is an indication that domain growth will eventually stop, thereby the system undergoes a microphase separation in this case.

19.6 Summary

The dynamics and morphology of multicomponent lipid membranes are investigated using large scale dissipative particle dynamics. The approach allows for the self-assembly of individual lipid particles into open lipid bilayer membranes or closed vesicles. The approach accounts for hydrodynamic interaction, and allows for the control of area-to-volume ratio. We found rich dynamics that depend strongly on the volume fraction of the lipids, line tension, and surface tension (or area-to-volume ratio). The dynamics is, in general, characterized by the formation of flat domains at early times and their coalescence following their Brownian motion. AT intermediate times, and in the presence of sufficient excess area, domains curve in order to decrease their interfacial energy. When line tension is high these curved domains may bud and vesiculate from the parent vesicle. However, the process of fission becomes prohibitive at low line tensions. At late times, domain growth of the remaining domains on the vesicle also proceeds through their Brownian motion. Domain growth of capped domains is characterized by different power law from that at earlier times, $L(t) \sim t^{-4/9}$.

Our model also allows for the investigation of the effect of the transbilayer asymmetry in the lipid composition. This is due to the absence of lipid flip-flops in the model. We found that the asymmetry in the lipid distribution lead to non-algebraic anomalously slow dynamics for moderate values of surface tension (or equivalently, area-to-volume ratio). The saturation in the average domain size in the case of transbilayer lipid distribution may be an explanation for the stabilization of lipid rafts in the plasma membranes of mammalian cells.

Achnowledgements

The authors would like the Petroleum Research Fund and the Indian Department of Science of Technology. Parts of the simulations were performed at the Danish Center for Scientific Computing. MEMPHYS is supported by the Danish National Research Foundation.

References

1. G. Gompper, M. Schick, *Self Assembling Amphiphilic Systems.* (Academic Press, New York, 1994).
2. B. Alberts, D. Bray, J. Lewis, M. Raff, K. Roberts, J.D. Watson, *Molecular Biology of the Cell.* 3rd edn. (Garland, New York, 1994).
3. S.J. Singer, G.L. Nicholson, Science **175**, 720 (1972).
4. K. Simmons, E. Ikonen, Nature **387**, 569 (1997).
5. M. Laradji, P.B. Sunil Kumar, Phys. Rev. Lett. **93**, 198105 (2004).
6. M. Laradji, P.B. Sunil Kumar, J. Chem. Phys. **123**, 224902 (2005).
7. M. Laradji, P.B. Sunil Kumar, submitted to Phys. Rev. E (2005).
8. T. Taniguchi, Phys. Rev. Lett. **76**, 4444 (1996).

9. P.B. Sunil Kumar, M. Rao, Phys. Rev. Lett. **80**, 2489 (1998).
10. P.B. Sunil Kumar, G. Gompper, R. Lipowsky, Phys. Rev. Lett. **86**, 3911 (2001).
11. P.J. Hoogerbrugge, J.M.V.A. Koelman, Europhys. Lett. **19**, 155 (1992).
12. P. Espagnol, Europhys. Lett. **40**, 631 (1997).
13. I.L. Lifshitz, V.V. Slyozov, J. Phys. Chem. Solids **19**, 35 (1962).
14. *Physics and the Architecture of Cell Membranes.* (Adam Hilger, Bristol, 1987).
15. A. Kusumi *et al.*, Traffic **5**, 213 (2004).
16. S. Mayor, M. Rao, Traffic **5**, 231 (2004).
17. S.L. Veatch, S.L. Keller, Phys. Rev. Lett. **94**, 148101 (2005).
18. *Structure and Dynamics of Membranes.* R. Lipowsky and E. Sackmann (Eds.) (Elsevier, Amsterdam, 1995).
19. C. Dietrich, L.A. Bagatolli, Z.N. Nolovyk, N.L. Thomson, M. Levi, K. Jakobsen, E. Gratton, Biophys. J. **80**, 1417 (2001).
20. S.L. Veatch, S.L. Keller, Phys. Rev. Lett. **89**, 268101 (2002).
21. T. Baumgart, S.T. Hess, W.W. Webb, Nature **425**, 821 (2003).
22. J. Bernandino de la Serna *et al.*, J. Biol. Chem. **279**, 40715 (2004).
23. J.M. Crane, L.K. Tamm, Biophys. J. **86**, 2965 (2004).

20

Solvent-Free Lipid-Bilayer Simulations: From Physics to Biology

M. Deserno

Max-Planck-Institut für Polymerforschung, Ackermannweg 10, 55128 Mainz, Germany

Abstract. Coarse-grained solvent-free simulation models enabling the study of self-assembling fluid lipid bilayers have been the goal of much recent modeling efforts, since their realization appeared to be quite intricate. This contribution reviews some of the challenges faced along the way, presents a surprisingly simple solution, and illustrates its capacity by means of three examples of biological interest.

20.1 Introduction

Fluid lipid membranes are self-assembled quasi-two-dimensional elastic objects with a vanishing transverse shear modulus. Their characteristic bilayer structure forms spontaneously in aqueous solution due to the amphiphilic nature of the constituent lipid molecules. The ability of fluid membranes to form complex structures [1], their mathematical challenges [2], and their ubiquity in biology [3,4] have always fascinated scientists from biology up to theoretical physics and mathematics.

One of their particularly fascinating features is that lipid bilayers display interesting physics on many different length scales. This ranges from lipid aggregation and packing behavior on the nanometer scale, solubilization of small or medium sized organic molecules or proteins, different bilayer phases, lateral stress profiles and their connection to elastic properties, and coupling of lipid composition to local membrane curvature, up to micron scale equilibrium structures, shape transitions, and fluctuation effects such as a length scale dependent renormalization of the bending modulus or repulsive forces stemming from thermal undulations. Naturally, such a host of diverse physical problems requires an equally broad array of tools for tackling them, as well as the knowledge of how to integrate insights gained by different means into a single picture – issues which apply to all scientists working within the trinity of experimental, theoretical and computational techniques. In this contribution we will have a closer look at certain simulational approaches to lipid bilayers.

Springer Proceedings in Physics, Volume 123
Computer Simulation Studies in Condensed-Matter Physics XIX
Eds.: D.P. Landau, S.P. Lewis and H.-B. Schüttler

The range of relevant length scales is closely mirrored by the range of bilayer models which have been devised to simulate lipid membranes on a computer. On the small scale we find *atomistic models* [5], in which individual lipid (and water) molecules are faithfully represented by properly connected constituent atoms resembling the chemical structure under study and interacting according to a well-tuned set of interactions (the "force field") which gives them their chemical identity. On the large scale membranes can be represented by *triangulated surfaces* [6] which are kept fluid by a dynamical change of the triangulation and whose energetics is chosen as the discrete counterpart of a continuum curvature elastic model. Between these extremes lies the wide field of *coarse grained* [7–17] membrane models which cover scales between a few ten up to a few hundred nanometers. In this range most of the specific chemical detail of lipid molecules becomes unimportant (except, of course, their amphiphilic nature), but the fact that bilayers are self-assembled from such constituents may not (yet) be entirely irrelevant. Consequently, the approach is to represent lipids (or groups of them) by more or less simplified toy-amphiphiles composed of generic "beads" linked by generic "springs", interacting with each other (as well as with generic "water") via some generic interactions that capture the amphiphilic nature of the problem. Clearly, without further careful parameter tuning specificity is not the strength of such models, but it is not their purpose either: They rather focus on universal physical questions (or those which are strongly believed to be universal), such as the coarsening behavior during lipid phase segregation, fusion kinetics, or composition-curvature-coupling. In this range, where (coming from the atomistic side) membranes start to be big enough in order to form three-dimensional curved structures, one particular aspect of a particle-based modeling matures from a mere nuisance to a severe obstacle that threatens to thwart further progress: The necessity to also incorporate explicit solvent particles. Why this is such a problem, why its avoidance seemed to be unexpectedly nontrivial, and how it can nevertheless be overcome in a surprisingly simple way is the topic of this contribution.

20.2 Solvent-free Membranes

Picture a spherical vesicle (i.e., a "bilayer bubble") of linear dimension L. The number of its constituent lipids will be proportional to its surface, i.e. proportional to L^2. However, holding the vesicle in the computer requires a simulation box with a linear dimension also proportional to L and a volume proportional to L^3, most of it filled with solvent molecules. The ratio between lipid and solvent molecules thus decreases as $1/L$; this way membrane physics becomes the study of a finite size effect. Yet, a little bit of thinking informs us that for many questions the fate of the solvent is of utter insignificance, and it seems like a waste of resources to devote so much effort to its explicit treatment. Yes, in the real world it is precisely the hydrophobic effect mediated by the solvent which drives the aggregation of lipid molecules in the first place; but there is no fundamental reason telling us that such an effective interaction could not be replaced by some (possibly generic) *effective potential.* In fact, the same unfavorable $1/L$ scaling haunts polymer simulations, and the insight to eliminate the solvent in favor of effective monomer-monomer interactions goes back a long way [18], without apparently having been picked up by the membrane community until rather recently. Why is that so?

20.2.1 The Solvent Strikes Back

Polymers are made by chemistry. Their long chains consist of monomers that are linked by chemical bonds – connections with an interaction energy of the order of electron volts, many times larger than the thermal energy, and thus their stability is unaffected by thermal fluctuations. In contrast, lipid membranes are made by physics. These self-assembled structures owe their existence to a delicate balance between energy and entropy. Their constituents are held together by the very same force which also threatens to tear them apart: thermal energy. Whether these forces are mediated by a surrounding solvent or whether they are represented by effective potentials – it is clear that their nature will be a fair bit more subtle than that of any old polymer bond, since both the structural stability of the bilayer as well as its subsequent material properties will depend on it.

When scientists first tried to eliminate the solvent from membrane simulations, they followed the same route that had proven to be so successful in the polymer field: To render the solvent poor for some set of monomers, one includes some attractive interaction between them. The canonical choice was the Lennard-Jones potential, the *Drosophila* of all generic molecular interaction potentials, and by tuning its depth ϵ the full range of solvent quality could be scanned. Unexpectedly, though, this did not work for lipid membranes. A largely unsuccessful (and hence unpublished) history of attempts showed that patient ϵ-tuning would not succeed: Either the cohesion was too weak, such that no self-assembly would take place, or it was too strong, such that an assembled two-dimensional aggregate would immediately go into the undesired *solid* (also termed "gel") phase. Even the less ambitious goal of stabilizing *pre-assembled* fluid membranes proved unattainable, thermal fluctuations would invariably breach their structural integrity. The subtle nature of aggregation forces turned out to be more than a mere academic issue and left the burning question of how to proceed next.

20.2.2 A New Hope

Effective interactions are generally of more complicated type than the ones between the underlying traced-out microscopic degrees of freedom. In the light of this fact, it appears to have been concluded that the hope to model lipid cohesion by simple pair forces was a bit bold. Hydrophobic forces may extend over a nanometer [19], well beyond the size of a single water molecule, and are believed to emerge from an intricate interplay between water enthalpy and entropy, details of which are disputed up to the present day [20]. Could it be that the above mentioned subtleties of lipid cohesion rest on aspects of the hydrophobic effect that are ill-represented by pair potentials?

In the early nineties Drouffe, Maggs, and Leibler [10] proposed a coarse grained solvent-free membrane model which incorporated lipid cohesion by multibody forces. It readily displayed self-assembly into a fluid bilayer phase. The key feature was a cohesive energy which depends on the density of surrounding hydrophobic units, but which *levels off* beyond some critical density. The following rational might have suggested this sub-linear behavior: One needs an increase in cohesive energy with density to drive aggregation, but to avoid the crystallization that has plagued the pair approaches it is cut short beyond some carefully chosen point. Even though the ansatz worked, it was not immediately developed further, and it took another decade before the idea was revived and successfully implemented in a modified lipid model by Noguchi and Takasu [11] and recently also by Wang and Frenkel [12].

Other suggestions have been made to circumvent a simple pair scenario, we'll mention only two: Farago [13] has suggested a model which rests on pair forces between the three different beads of a coarse-grained lipid with those of a neighboring one; however, it employs six *different* potentials whose functional form, attraction range, depth, and excluded volume were optimized in a rather lengthy "trial and error" process. Two notable guiding principles were the incorporation of a rather shallow attractive minimum to increase lipid mobility and a non-additive repulsion to optimize the hydrophobic constraint and maybe also prevent crystallization. Brannigan et al. [14] have proposed a model in which lipids are represented by spherocylinders and in which the interaction between them is also a function of their relative *angle*. Next to the tail-tail attraction the alignment strength becomes a second parameter, which suitably chosen enables fluid bilayer phases. While both models have certain difficulties, specifically with respect to self-assembly, they provide a strong hint that avoiding the pair level is not necessary after all.

20.2.3 The Return of the Pair Force

While it proved to be *possible* to obtain solvent-free self assembly on the basis of multibody potentials or other more or less sophisticated techniques, the question remained whether these means are *desirable*, i.e., worth the price to pay. The pertinent unease fed on the usual technical and conceptual problems and unfamiliarities linked to anything which exceeds a point-pair-level, such as the definition of global and local stresses, the difficulties to interpret model ("tuning") parameters, the need to calculate torques and propagate angular momenta (at least within molecular dynamics), and simply code-efficiency. However, the experience gained with the models of Farago [13] and Brannigan et al. [14] strongly suggested that a simple pair solution must exist.

In 2005 two groups independently presented solvent-free coarse grained bilayer models which rested exclusively on simple pair potentials and displayed an unassisted self-assembly into a fluid bilayer phase. One was proposed by Brannigan et al. [15], the other one by Cooke et al. [16]. Even though these two models differed in several respects, they shared one apparently crucial feature: The cohesive potential mediating the hydrophobic aggregation displays a range of attraction which exceeds the typical width of a Lennard-Jones potential[1]. In fact, Cooke and Deserno [17] subsequently showed that slightly extending the Lennard-Jones range by "adding" a flat piece at the minimum readily enables fluid self-assembly in a way which the original potential refused to do.

In retrospect it is of course easy to rationalize why broadening the range is the method of choice. Farago [13] already pointed out that mutual lipid mobility ought to be aimed for, even though it remained unclear whether this aim is better pursuit by a *long range* or by *soft repulsions.* Probably one of the strongest hints should have been the knowledge from the colloidal community that interaction potentials whose range of attraction is too short compared to their excluded volume size do not give rise to a gas-liquid phase transition (i.e., they do not have a distinct dense fluid phase) [21]. In the three-dimensional case a 6–12 Lennard-Jones potential suffices, yet in two dimensions – with the additional complication of enforcing a condensed phase in the first place – there is no reason to believe that this will also work.

Both the (second) Brannigan [15] and the Cooke [16, 17] model offer a solution to the quest for a solvent-free coarse-grained membrane model. They are still very "young" and must prove their usefulness by application to real world problems. In the remainder of this article I will illustrate the Cooke model and its application to a few typical meso-scale problems. This selection is based on my own research experience, accessibility to examples and data, and (undeniably) my own involvement in the latter model. The reader (and Brannigan et al.) may forgive this personal bias.

20.3 The Cooke Model

20.3.1 Definition

The coarse grained lipids are represented by three linearly linked beads, one end bead representing the hydrophilic head, the two other ones the tail. Their size is fixed via a Lennard-Jones potential truncated at the minimum:

$$V_{\mathrm{rep}}(r;b) = \begin{cases} 4\epsilon\left[\left(\frac{b}{r}\right)^{12} - \left(\frac{b}{r}\right)^{6} + \frac{1}{4}\right] , & r \le r_{\mathrm{c}} \\ 0 , & r > r_{\mathrm{c}} \end{cases} , \tag{20.1}$$

[1] In the Brannigan model this feature is restricted to special "interface" beads between the hydrophilic heads and the hydrophobic tails. Between the latter there is still an attraction, but it is of "standard" short range type.

where $r_c = 2^{1/6}b$ and ϵ is the unit of energy. The aspect ratio of the lipid can be influenced by changing the size of the head compared to the tail beads [22]. The choice $b_{\text{head,head}} = b_{\text{head,tail}} = 0.95\,\sigma$ and $b_{\text{tail,tail}} = \sigma$ leads to basically cylindrical lipids which will yield bilayer phases (as opposed to spherical or cylindrical micelles); σ will be used as the unit of length.

The three beads are linked by two FENE bonds

$$V_{\text{bond}}(r) = -\frac{1}{2}k_{\text{bond}}\, r_\infty^2 \log\left[1 - (r/r_\infty)^2\right] , \tag{20.2}$$

with stiffness $k_{\text{bond}} = 30\,\epsilon/\sigma^2$ and divergence length $r_\infty = 1.5\,\sigma$, standard values in the field of polymer simulations. In order to straighten the lipid one would usually add a bending potential of the form $\frac{1}{2}K(\pi - \vartheta)^2$, where ϑ is the angle formed by the three beads and K the modulus tuning the bending stiffness. This three-body-term can be avoided by reverting to a "poor man's bending potential", namely, a harmonic spring with rest length 4σ between head-bead and second tail-bead

$$V_{\text{bend}}(r) = \frac{1}{2}k_{\text{bend}}(r - 4\sigma)^2 . \tag{20.3}$$

In lowest order this is equivalent to the three body bending term if one sets $K = k_{\text{bend}}\sigma^2$. The Cooke model uses $k_{\text{bend}} = 10\,\epsilon/\sigma^2$, which for $k_B T = \epsilon$ gives angular fluctuations of the order of $\pm 20°$.

The absence of explicit solvent molecules and the hydrophobic effect they would give rise to is compensated by an attractive interaction between all tail beads. This is achieved by the potential

$$V_{\cos}(r) = \begin{cases} -\epsilon & , \quad r < r_c \\ -\epsilon\,\cos^2\dfrac{\pi(r - r_c)}{2w_c} & , \quad r_c \le r \le r_c + w_c \\ 0 & , \quad r > r_c + w_c \end{cases} ; \tag{20.4}$$

for $r > r_c$ its attractive strength of ϵ smoothly tapers to zero over a range of width w_c. The precise form in which this happens is fairly irrelevant [17], what matters is that tuning w_c can create potentials which extend beyond the usual Lennard-Jones range. It must be understood that this does *not* imply potentials which are "long ranged" in the sense of "computationally expensive". The value $w_c = 1.6\,\sigma$, used below for most illustrations, yields a potential that is zero for $r/\sigma \ge 2^{1/6} + 1.6 \approx 2.7$, only 8% further out than the frequently used Lennard-Jones cutoff $r/\sigma = 2.5$.

Thermal properties of this model may be probed using a variety of different techniques. The choice followed below is Molecular Dynamics (MD) simulations, for which the `ESPResSo` package [23] has been used. The canonical state was reached by using a Langevin thermostat [24] (time step $\delta t = 0.01\,\tau$ and a friction constant $\Gamma = \tau^{-1}$ in Lennard-Jones units [25]). Simulations were performed using a cuboid box with sides $L_x = L_y$, L_z (subject to periodic

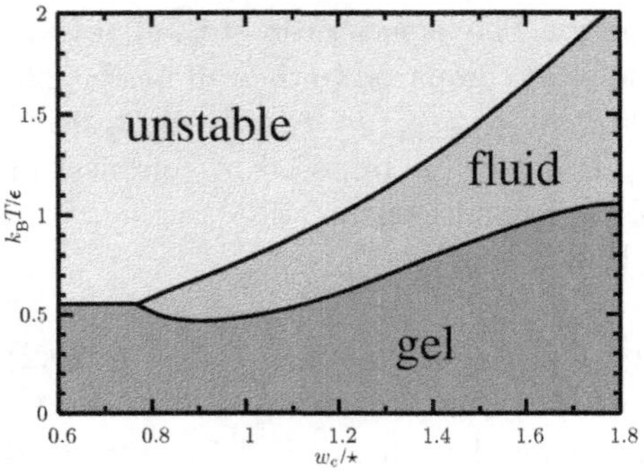

Fig. 20.1. Simplified phase diagram of the Cooke model in the plane of potential width w_c and temperature at zero lateral tension. Observe that only for $w_c \gtrsim 0.8\,\sigma$ a temperature window opens up within which fluid membranes exist

boundary conditions) either at fixed box size or under conditions of constant lateral tension, which were implemented via a modified Andersen barostat [26] allowing box resizing in x and y dimensions only (with a box friction $\Gamma_{\text{box}} = 2 \times 10^{-4}\,\tau^{-1}$ and box mass within the range $Q = 10^{-5} \ldots 10^{-4}$).

After having defined the model, let us have a look at the self-assembly and material properties of the resulting membranes.

20.3.2 Self Assembly and Overall Phase Behavior

Provided the strength ϵ of the pair cohesion (20.4) is large enough compared to the thermal energy k_BT, a random distribution ("gas") of lipids will spontaneously assemble into aggregates displaying a local bilayer structure. However, a small fraction of lipids (typically about 1% for fluid membranes not too close to the evaporation transition) remains free, establishing a chemical equilibrium between aggregated and "stray" lipids. The global topology of the aggregate depends on further details. It may zip up across the periodic boundary conditions to form a flat bilayer spanning the box. If the box is too big for that, pancake-like aggregates will form which beyond a critical size (proportional to the ratio of bending modulus and line energy [27]) buckle under their own line tension and close up, yielding a vesicle.

To study flat membranes more closely, box-spanning bilayers are very suitable, and to render different simulations comparable, it is convenient to subject them to a fixed tension (zero in the present case). Figure 20.1 shows the result of scanning two key parameters: interaction range w_c and the ratio k_BT/ϵ of thermal energy and potential depth. At too large temperature bilayers, even though pre-assembled, fall apart. At sufficiently small temperatures they solidify[2] (visible in a variety of observables, such as order parameters or diffusion

[2] More precisely, the low temperature part of lipid phase diagrams (including this one) subdivides into a variety of distinct phases, generically referred to as "solid" or "gel", but none of them is actually a true solid (which strictly speaking does not exist in two dimensions).

constant). And between the gas and the solid phases we find a region of fluid membranes, *provided* the range w_c is big enough.

20.3.3 Material Properties in the Fluid Phase

While all membranes in the intermediate temperature region in Fig. 20.1 share the property of being fluid, their other material properties vary over the w_c-T-plane. A detailed study can be found in [17]; here we only highlight the major points.

Bending Rigidity. The central large-scale material parameter of a fluid membrane is its *bending rigidity*, the modulus κ describing the energy per unit area to assume a certain local extrinsic curvature K (which is the sum of the two principal curvatures) [28]:

$$e_{\text{bend}} = \frac{1}{2}\kappa K^2 \ . \tag{20.5}$$

For essentially flat membranes $K \simeq \Delta h(x, y)$, where $h(x, y)$ describes the height of the membrane above some reference plane and Δ is the Laplacian in this plane. If the membrane is spanned in a quadratic frame of area L^2, its shape can be expanded in Fourier modes. The total energy – the integral of (20.5) over the membrane – is then a sum of independent Fourier modes, each entering quadratically. Using the equipartition theorem one then finds that at zero lateral tension, the situation realized in the simulation, the modes fluctuate like

$$\langle |h_q^2| \rangle = \frac{k_B T}{L^2 \kappa q^4} \propto q^{-4} \ . \tag{20.6}$$

The fluctuation spectrum thus gives access to the bending modulus. For the Cooke model, one finds values between about $3\,k_B T$ and $30\,k_B T$ – exactly within the experimentally relevant regime for phospholipid bilayers [29].

Bilayer Order. One simple measure for bilayer order is the *orientational order parameter* $S = \frac{1}{2}\langle 3(\boldsymbol{a}_i \cdot \boldsymbol{n})^2 - 1 \rangle_i$, where $\boldsymbol{a}_i$ is the orientation of lipid i and $\boldsymbol{n}$ is the local bilayer normal at that position (which can often reasonably well be approximated by the average bilayer normal). In the case of perfect alignment of the lipids with the bilayer normal, we would have $S = 1$, no correlation whatsoever would correspond to $S = 0$. One finds values roughly between 0.4 and 0.6 for bilayers close to the evaporation boundary, while bilayers close to the gel transition have values around 0.8. This order parameter correlates well with other bilayer properties, such as, for instance, the overlap between the two monolayers (larger S corresponds to less overlap), the area per lipid (larger S means a smaller area per lipid) or the bending modulus (larger S means stiffer membranes). All these observables change discontinuously across the gel transition, visible also in the appearance of hysteresis. The gel phases subdivides into further phases which can be distinguished using further order parameters, but this will not be discussed here.

Dynamical Properties. One of the crucial dynamical observables is the lipid diffusion constant. It can be obtained from the mean squared displacement of lipids, but one has to convince oneself first that the small but fast contribution of stray lipids (which obviously translate much faster) will not significantly endanger the "naive" measurement. One finds that the diffusion constant varies about one order of magnitude inside the fluid region and decreases when going from the evaporation boundary towards the gel boundary, where upon gelling it drops suddenly by about two orders of magnitude (which is in good agreement with experiment [30]). A typical value for the diffusion constant in the fluid phase is about $0.01\,\sigma^2/\tau$ (where τ is the Lennard-Jones time).

A second interesting dynamical observable is the flip-flop-rate r_{f}. It measures the frequency with which lipids on average flip between monolayers, i.e., $1/r_{\mathrm{f}}$ is the average time between flips. The flip-flop-rate varies over almost three orders of magnitude inside the fluid region ($10^{-6}...10^{-3}\,\tau^{-1}$), larger values again corresponding to more disordered membranes close to the evaporation boundary. These values are very large compared to real lipid bilayers. They imply that lipids typically diffuse a distance corresponding to several times the bilayer diameter before flipping, which is short compared to real systems (where flip-flop times may be hours). This is a direct consequence of the simplicity of the lipids (extending the number of beads per lipid would reduce the flip-flop-rate) and limits the interpretation of dynamical results (as does of course also the absence of the solvent in the first place). However, from a static point of view this rapid flipping can be seen as an advantage, because it implies that the two sides of a bilayer achieve an equilibrium composition rather quickly (as might be needed when starting with a two-component mixed membrane which is then subjected to a deformation, after which one waits for equilibration).

Length and Time Scale Mapping. There are two obvious ways in which one could translate the length scale σ into a real length. The first proceeds via the area per lipid, which for typical phospholipid membranes has values around $0.75\,\mathrm{nm}^2$, while we find values between 1.1 and $1.5\,\sigma^2$, suggesting $\sigma \simeq 0.7$–$0.8\,\mathrm{nm}$ if one assumes that one coarse grained lipid in fact corresponds to one real lipid. Another possibility is to compare the bilayer thickness, which in real life is between 4 and 5 nm, while we find about $5\,\sigma$, giving $\sigma \simeq 0.8 - 1\,\mathrm{nm}$. The close agreement between these two mappings also illustrates that the *aspect ratio* of our lipids is actually quite close to the one for real lipids.

Mapping of time scales typically proceeds via the diffusion constant. Since a typical value in our case is $0.01\,\sigma^2/\tau$, while real membranes have about $1\,\mu\mathrm{m}^2/s$ [30], this gives $\tau \simeq 3 - 10\,\mathrm{ns}$. These estimates are of course not very quantitative, but they illustrate that it is possible to reach the millisecond regime with these membranes, since several $10\,000\,\tau$ can be reached with several thousand lipids without much difficulty. It must be noted, however, that the dynamical discrepancy between diffusion constant and flip-flop rates as mentioned above advises caution when assessing these numbers.

20.4 Applications

In order to illustrate the model described in the previous section, it will be applied here to three situations which all look at the physics of lipid mixtures. The first two examples focus on the demixing of two lipid species which mutually dislike each other, the third example asks the question whether the shape of a lipid might have an influence on its positioning in the membrane.

20.4.1 Domain Induced Budding

The vast majority of work about lipid membranes (be it experimental, theoretical, or computational) has been devoted to one component bilayers, consisting of only one species of lipid molecules. This focus on a simple situation is understandable (and there exists of course much interesting physics), but it misses biological reality by about two orders of magnitude. Cells resort to a *few hundred* different lipids to build their various membrane structures (which also differ noticeably in their composition) [31] – not to mention the large number of trans-membrane proteins solubilized in the bilayer.

There are two approaches to this difficulty. One is to hope that in a mixture consisting of so many components the mixing entropy is so large that under many (typical?) conditions everything ought to be homogeneously mixed. Then the membrane appears on sufficiently large length scales as a one-component system. The second approach is to humbly remember that in order to get from 1 to 100, one first has to pass 2 – and then study binary (or maybe ternary) systems. Here we would like to discuss one of the simplest effects occurring upon binary fluid-fluid demixing: domain induced budding.

Picture a flat two-component membrane holding a phase-separated domain of some area $A_{\rm d}$. It is distinguishable as an entity due to its surrounding contact line, which comes at the price of a line tension γ. The latter also renders the patch roughly circular with a radius $R_{\rm d} \simeq \sqrt{A_{\rm d}/\pi}$, since this minimizes the line energy $2\pi R_{\rm d}\gamma$. Still, this energy grows with growing patch size. Lipowsky has pointed out [32] that beyond a critical size the domain prefers to contract the phase line at the expense of *buckling* into the third dimension. Considering only bending[3], the scale invariant Helfrich Hamiltonian (20.5) renders the situation very transparent: any spherically budded region costs curvature energy $8\pi\kappa$ independent of the domain size.[4] Balancing this against the line energy yields budded vesicles with a radius $R_{\rm v} \simeq 2\kappa/\gamma$. The situation resembles the close-up transition from "pancake" membranes patches (created e.g. by ultrasonication) to vesicles [27], but since in the present case the line energy is much smaller, the size of the created vesicles is bigger (tension would also help here), such that they can be visualized *optically* [33].

[3] Additional lateral membrane tension complicates the situation only slightly [32].

[4] This requires both separate phases to remain fluid, which typically means that at least one more component is present as a "fluidizer", usually cholesterol.

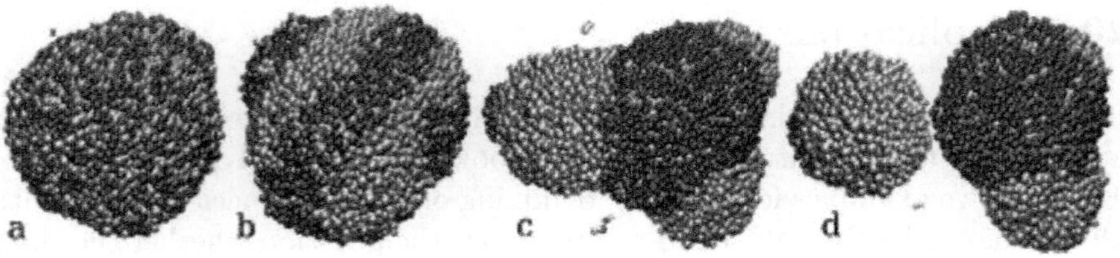

Fig. 20.2. Phase separation and budding sequence for a mixed vesicle pre-assembled from 2000 A-lipids and 2000 B-lipids. After an equilibration time of $4000\,\tau$, during which $w_c^{AA} = w_c^{BB} = w_c^{AB} = 1.5\epsilon$ (i.e., all lipids identical), the cross term was reduced to $w_c^{AB} = 1.3\epsilon$. The snapshots correspond to the following waiting times after this reduction step: (**a**) $0\,\tau$; (**b**) $1000\,\tau$; (**c**) $5500\,\tau$; (**d**) $10900\,\tau$

This scenario has been checked recently using triangulated membranes [34] and dissipative particle dynamics (DPD) simulations [8,9]. While the former have no explicit solvent (but also no explicit lipids, such that domains must be artificially postulated), the latter simulations painfully notice the overwhelming dominance of solvent degrees of freedom discussed at the beginning of Sect. 20.2. Solvent free models, in contrast, do not have this difficulty. The sequence of snapshots in Fig. 20.2 shows the split-up of a vesicle composed of a 50:50 mixture of mutually incompatible lipids using the Cooke model. Such a simulation takes only a few days on a current single processor machine.

20.4.2 Domain Coarsening

All domains start small. How do they grow? This may be quantified for instance by measuring their number N_d as a function of time t. Since an initial random distribution of lipids translates into an exponential domain size distribution in which the small ones vanish first, the number of domains first drops exponentially. But soon afterwards it enters a power law regime in which $N_d \propto t^{-\alpha}$. The long time behavior of the present solvent free model, using a Langevin thermostat [24], then corresponds to a kind of "Rouse model' for membranes, i.e., a situation dominated by thermal noise. In [16], an exponent $\alpha \simeq 0.38$ is measured. This can be rationalized by the following scaling argument due to Binder and Stauffer [35]: Each patch moves by detachment and attachment of individual lipids surrounding its surface like a "halo". A single such process then moves the patch by an amount proportional to L^{-d}, where L is its current size and d its spatial dimension. This happens everywhere on the surface, so the entire motion is diffusive with a length scale dependent diffusion constant $D(L) \propto L^{d-1} \times (L^{-d})^2 = L^{-d-1}$. Since halos are tightly bound to the patch, coarsening happens via patch coalescence, whereby the patch volume increases by the size of one patch, $\Delta V = L^d$. The time Δt one has to wait for such an event to occur is given by the time it takes to diffuse the distance between patches. There is only one length scale in the system, so this distance is also proportional to L. We then find

$$L^{d-1}\frac{\mathrm{d}L}{\mathrm{d}t} \sim \frac{\mathrm{d}V}{\mathrm{d}t} \approx \frac{\Delta V}{\Delta t} \sim \frac{L^d}{L^2/D(L)} \sim L^{-3} \,. \tag{20.7}$$

This yields the differential scaling relation $\mathrm{d}L/\mathrm{d}t \sim L^{-d-2}$, giving $L \propto t^{1/(3+d)}$. Since finally the number of domains N_d times the domain volume L^d must remain constant, we find $N_\mathrm{d} \sim t^{-d/(d+3)}$, which for $d = 2$ gives the scaling exponent $\alpha = 0.4$, close to the result of the simulation.

Being a dynamical question, the absence of explicit solvent may matter – and in fact it does, for two reasons. First, the solvent enclosed by the vesicle provides a *volume constraint* which might stop or slow down the formation of buds. Second, hydrodynamic effects influence the friction of patches and thus their diffusion. Both points are carefully studied in [9].

20.4.3 Composition-Curvature-Coupling

Creation of small vesicles from larger membranes is a very frequent process inside living cells, since vesicles are the "shuttles" used for active transport between different cellular compartments [4]. One of the obvious mysteries here is why the delicately adjusted different lipid compositions of different compartments are not quickly washed out by this vesicle trafficking, which is of course also *lipid* trafficking. A bold answer is that this very trafficking is one of the means for *sorting* lipids [36]. One hypothesis for how this might be physically accomplished involves the coupling of individual lipids to the curvature of the membrane they reside in [37,38]. Transport vesicles are much more strongly curved then the membranes they originate from, and during their formation even more highly curved intermediate structures occur (such as budding necks or fission stalks). Would it be conceivable that these strongly curved membranes preferentially incorporate or exclude lipids based on their molecular *shape*?

Quantitatively checking this hypothesis in a simulation requires one to treat sufficiently many lipids (thus permitting bilayer curvature) over a sufficiently long time (thus enabling equilibrium to be established), for which coarse grained models are ideally suited. In the present model a simple way to tune lipid shape is to vary the diameter of the lipid head bead, which indeed yields the entire sequence of surfactant morphologies (spherical micelles, cylindrical micelles, bilayers) in a way that can be quantitatively matched with the packing parameter [22]. The hypothesis can then, for instance, be tested by measuring the distribution of differently shaped lipids between inner and outer monolayer of vesicles, for which the sign of the extrinsic curvature is different [22]. But before doing this, it is advisable to first develop a simple model in order to understand what to expect from such a simulation.

Let there be M lipids in total, M_o in the outer monolayer and M_i in the inner one of the vesicle. Let there also be N special lipids which are curvature-sensitive (see below), N_o in the outer and N_i in the inner monolayer. The radius of the vesicle is identified as the radius towards the bilayer midplane

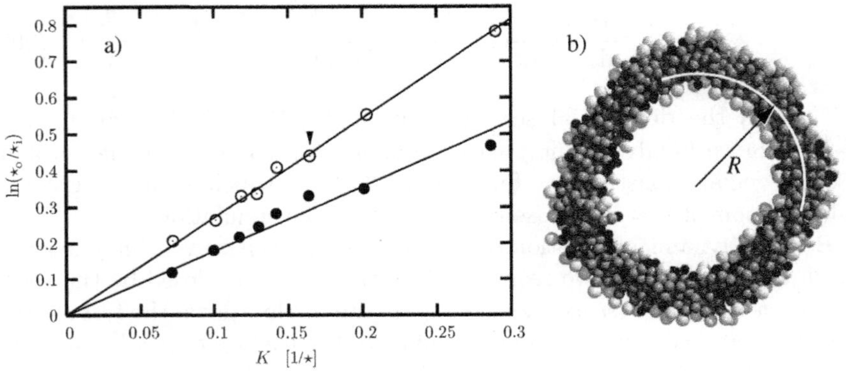

Fig. 20.3. (**a**) Logarithm of interleaflet lipid sorting ratio as a function of vesicle curvature K for two different lipid asymmetries. Filled circles correspond to special lipids with head size $b_{\mathrm{head,head}} = 1.1\,\sigma$, while open circles correspond to the bigger lipid curvature $b_{\mathrm{head,head}} = 1.2\,\sigma$. The lines are fits to (20.9). (**b**) Cross-section through a vesicle corresponding to the point indicated in (**a**) by an arrow. White heads have $b_{\mathrm{head,head}} = 1.2\sigma$, black heads have $b_{\mathrm{head,head}} = 0.8\sigma$

and it is denoted by R. The corresponding curvature is $K = 2/R$. We will assume that the entropy S of the special lipids embedded in the normal ones can be written as an ideal lattice gas and that their energy E of coupling to curvature is harmonic with modulus $\mathcal{M}$. Then, the free energy is

$$F = E - TS = \sum_{x\in\{\mathrm{o,i}\}} N_x \left[\frac{1}{2}\mathcal{M}\,(K_x - K_\ell)^2 + k_{\mathrm{B}}T\,(\ln\phi_x - 1)\right] , \qquad (20.8)$$

where K_{o} and K_{i} are the curvatures of the outer and inner monolayer, respectively (in the simplest approximation they are $2/R$ and $-2/R$), K_ℓ measures the intrinsic "lipid curvature", and where we have also introduced the special lipids fractions $\phi_x = N_x/M_x$. The free energy F is now minimized with respect to both N_x, but subject to the constraint that $N = N_{\mathrm{o}} + N_{\mathrm{i}}$ (incorporated by a Lagrange multiplier λ). Differentiation then gives $\ln\phi_{\mathrm{x}} = \lambda - \frac{1}{2}\beta\mathcal{M}(K_{\mathrm{x}} - K_\ell)^2$, which can be rewritten as [22]

$$\ln\frac{\phi_{\mathrm{o}}}{\phi_{\mathrm{i}}} = 2\beta\mathcal{M}K_\ell\,K\,. \qquad (20.9)$$

The logarithm of the sorting ratio is proportional to the curvature of the vesicle, with a prefactor indicating the strength of coupling.[5]

[5] The more accurate expressions $K_{\mathrm{o}} = 2/(R+\delta)$ and $K_{\mathrm{i}} = 2/(R-\delta)$, where δ is something like half a monolayer diameter, give a correction term $\beta\mathcal{M}K^3\delta$. With its help $\mathcal{M}$ and K_ℓ could be determined separately, but a quick glance at Fig. 20.3 shows that the data are not good enough to extract this cubic correction.

The idea would thus be to simulate vesicles of different size and plot the logarithm of the sorting ratio as a function of vesicle curvature. In [22] vesicles have been simulated consisting of between 1000 and 16 000 lipids (corresponding to radii between $5\,\sigma$ and $30\,\sigma$), which themselves were a 50:50 mixture of large-headed and small headed lipids with an *overall* vanishing effect on the curvature. The result for the enhancement factor for the big-headed lipids is shown in Fig. 20.3 for two different strengths of lipid curvature. Apparently, (20.9) describes the data very well. Assuming $\sigma \simeq 1\,\mathrm{nm}$ implies a coupling strength of $\mathcal{M}K_\ell \approx 3.6\,\mathrm{pN\,nm}^2$ for the weaker curved lipid and $\mathcal{M}K_\ell \approx 5.6\,\mathrm{pN\,nm}^2$ for the stronger one.

The effect is clearly measurable, but it is quite small. Even highly curved transport vesicles ($R \approx 50\,\mathrm{nm}$) only imply a sorting efficiency of a few percent, not the experimentally observed 95% [37]. One might suggest the use of even more highly curved lipids, but these no longer prefer to be in a bilayer shape and rather aggregate into cylindrical or spherical micelles [22]. It is much more likely that in nature the coupling between lipids and curvature works in concert with other mechanisms, such as a lipid phase segregation, in which case curvature coupling is only one of several effects which tip the balance towards selectively incorporating special lipids into buds.

Acknowledgments

It is a pleasure for me to thank Ira Cooke for his many inspire contributions to the research presented here. I am also grateful to Oded Farago, Vagelis Harmandaris, Gregoria Illya, Kurt Kremer, Olaf Lenz, Bernward Mann, Martin Müller, Hiroshi Noguchi, Benedict Reynolds, Friederike Schmid, and Eva Sinner for many useful discussions. I finally would like to acknowledge financial support by the German Science Foundation under grant No. De775/1-3.

References

1. W.M. Gelbart, A. Ben-Shaul, D. Roux (Eds.), *Micelles, Membranes, Microemulsions, and Monolayers.* (Springer, New York 1994).
2. D.R. Nelson, T. Piran, S. Weinberg (Eds.), *Statistical Mechanics of Membranes and Surfaces*, 2nd edn. (World Scientific, Singapore 2004).
3. R. Lipowsky, E. Sackmann (Eds.), *Structure and Dynamics of Membranes – From Cells to Vesicles*, vol. 1 of the *Handbook of Biological Physics* (Elsevier, Amsterdam 1995).
4. H. Lodish, A. Berk, S.L. Zipursky, P. Matsudaira, D. Baltimore, J. Darnell, *Molecular Cell Biology.* 2nd edn. (W.H. Freeman, New York, 2000).
5. D.P. Tieleman, S.J. Marrink, H.J.C. Berendsen, Biochim. Biophys. Acta **1331**, 235 (1997); S.E. Feller, Curr. Opin. Colloid Interface Sci. **5**, 217 (2000); L. Saiz, S. Bandyopadhyay, M.L. Klein, Biosci. Rep. **22**, 151 (2002).
6. G. Gompper, D.M. Kroll, Chap. 12 in [2].

7. R. Goetz, R. Lipowsky, J. Chem. Phys. **108**, 7397 (1998); R. Goetz, G. Gompper, R. Lipowsky, Phys. Rev. Lett. **82**, 221 (1999); J.C. Shelley, M.Y. Shelley, R.C. Reeder, S. Bandyopadhyay, M.L. Klein, J. Phys. Chem. B **105**, 4464 (2001); R.D. Groot, K.L. Rabone, Biophys. J. **81**, 725 (2001); J. Shillcock, R. Lipowsky, J. Chem. Phys. **117**, 5048 (2002); S.J. Marrink, A.E. Mark, J. Am. Chem. Soc. **125**, 11144 (2003); M.J. Stevens, J. Chem. Phys. **121**, 11942 (2004); R. Chang, G.S. Ayton, G.A. Voth, J. Chem. Phys. **122**, 244716 (2005).
8. S. Yamamoto, S. Hyodo, J. Chem. Phys. **118**, 7937 (2003).
9. M. Laradji, P.B. Sunil Kumar, Phys. Rev. Lett. **93**, 198105 (2004).
10. J.-M. Drouffe, A.C. Maggs, S. Leibler, Science **254**, 1353 (1991).
11. H. Noguchi, M. Takasu, Phys. Rev. E **64**, 041913 (2001); J. Chem. Phys. **115**, 9547 (2001); Biophys. J. **83**, 299 (2002); Phys. Rev. E **65**, 051907 (2002).
12. Z.-J. Wang, D. Frenkel, J. Chem. Phys. **122**, 234711 (2005).
13. O. Farago, J. Chem. Phys. **119**, 596 (2003).
14. G. Brannigan, F.L.H. Brown, J. Chem. Phys. **120**, 1059 (2004); G. Brannigan, A.C. Tamboli, F.L.H. Brown, J. Chem. Phys. **121**, 3259 (2004).
15. G. Brannigan, P.F. Philips, F.L.H. Brown, Phys. Rev. E **72**, 011915 (2005).
16. I.R. Cooke, K. Kremer, M. Deserno, Phys. Rev. E **72**, 011506 (2005).
17. I.R. Cooke, M. Deserno, J. Chem. Phys. **123**, 224710 (2005).
18. K. Binder (Ed.), *Monte Carlo and Molecular Dynamics Simulations in Polymer Science.* (Oxford University Press, Oxford, 1995).
19. J.N. Israelachvili: *Intermolecular and Surface Forces.* 2nd edn. (Academic Press, London, San Diego, 1992).
20. N.T. Southall, K.A. Dill, A.D.J. Haymet, J. Phys. Chem. B **106**, 521 (2002).
21. A.P. Gast, C.K. Hall, W.B. Russel, J. Coll. Interface Sci. **96**, 251 (1983); M.H.J. Hagen, D. Frenkel, J. Chem. Phys. **101**, 4093 (1994); A.A. Louis, Phil. Trans. R. Soc. Lond. A **359**, 939 (2001).
22. I.R. Cooke, M. Deserno, Biophys. J. (submitted).
23. A. Arnold, B.A.F. Mann, H.J. Limbach, C. Holm, Comp. Phys. Comm. (in press); see also: `http://www.espresso.mpg.de`.
24. G.S. Grest, K. Kremer, Phys. Rev. A. **33**, 3628 (1986).
25. M.P. Allen, D.J. Tildesley: *Computer Simulation of Liquids.* (Clarendon, Oxford, 1990).
26. A. Kolb, B. Dünweg, J. Chem. Phys. **111**, 4453 (1999).
27. W. Helfrich, Phys. Lett. A **50**, 115 (1974).
28. P.B. Canham, J. Theoret. Biol. **26**, 61 (1970); W. Helfrich, Z. Naturforsch. **28c**, 693 (1973).
29. U. Seifert, R. Lipowsky: Chap. 8 in [3].
30. P.F. Fahey, W.W. Webb, Biochem. **17**, 3046 (1978).
31. E. Sackmann: Chap. 1 in [3].
32. R. Lipowsky, Biophys. J. **64**, 1133 (1993).
33. T. Baumgart, S.T. Hess, W.W. Webb, Nature **425**, 821 (2003).
34. P.B. Sunil Kumar, G. Gompper, R. Lipowsky, Phys. Rev. Lett. **86**, 3911 (2001).
35. K. Binder, D. Stauffer, Phys. Rev. Lett. **33**, 1006 (1974).
36. K. Simons, G. van Meer, Biochem. **27**, 6197 (1988).
37. S. Mukherjee, T.T. Soe, F.R. Maxfield, J. Cell. Biol. **144**, 1271 (1999).
38. H.T. McMahon, J.L. Gallop, Nature **438**, 590 (2005).

21

Computer Simulation of Models for Confined Two-Dimensional Colloidal Crystals: Evidence for the Lack of Positional Long Range Order

A. Ricci[1], P. Nielaba[2], S. Sengupta[3], and K. Binder[1]

[1] Institut für Physik, Johannes Gutenberg-Universität Mainz, 55099 Mainz, Germany
[2] Physics Department, University of Konstanz, 78457 Konstanz, Germany
[3] S.N. Bose National Centre of Basic Sciences, Salt Lake, Calcutta 700098, India

Abstract. Point particles interacting with inverse power-law potentials can describe colloidal particles at the air-water interface, a model system for the experimental study of melting in two dimensional ($2d$). Monte Carlo simulations are used to investigate the effect of confinement on such a system in the crystalline state. It is shown that the state of the system (a strip of width D) depends very sensitively on the precise boundary conditions at the two parallel "walls" providing the confinement. In the y-direction parallel to these walls, orientational order is locally always enhanced but positional long range order is destroyed, if the wall is just a smooth repulsive boundary with no corrugation in the y-direction. Then the mean-square displacement of two particles n lattice parameters apart in y-direction cross over from the logarithmic increase (characteristic for $2d$) to a linear increase (characteristic for $1d$). However, using a suitable corrugated boundary stabilizes the positional long range order, illustrating the fact systems which may exhibit long range order are sensitive to boundary conditions at their surfaces even if these boundaries are far apart.

21.1 Introduction and the Choice of the Model

Using suitable colloidal particles which have sizes in the μm range it has become possible to study fluids and solids in $2d$ experimentally, confining these particles at the air-water interface [1]. Using laser fields also confinement of single rows of particles ($1d$) is possible [2]. Since the ordering of the particles can be directly observed in real space by video microscope techniques, fascinating concepts on the statistical mechanics of low-dimensional systems [3–5] may now be tested experimentally.

However, the experiments can infer the precise functional form of the interparticle potential at best indirectly, if at all, and hence it is very useful

Springer Proceedings in Physics, Volume 123
Computer Simulation Studies in Condensed-Matter Physics XIX
Eds.: D.P. Landau, S.P. Lewis and H.-B. Schüttler

to provide guidance to the interpretation of experiments by computer simulation of suitable model systems. In this spirit, we present a simulation of a system of point particles interacting with a repulsive power law potential, $V(r) = \varepsilon(\sigma/r)^{-n}$, focusing on systems that are *in between these dimensionalities* $d = 2$ *and* $d = 1$, namely particles confined in strips of a finite width D [6]. The parameters ε and σ set the scales for the strength and range of the potential and can be chosen as $\varepsilon = 1$, $\sigma = 1$. While for the experimental system [1] it would be most realistic to choose $n = 3$, from a computational point of view it is more convenient to choose a faster decaying potential, $n = 12$. We emphasize that the phenomena which we discuss should be universal, and not dependent on details of the potential. The above potential can be cut off at $r_c = 5.0$. Choosing a temperature $T = 1 (k_B \equiv 1)$ and a density $\rho = 1.05$, a bulk $2d$ system would be deep in the crystalline phase, since for the chosen density melting occurs at [7] $T \approx 1.35$.

We choose two types of wall potentials which provide the confinement in a strip with walls oriented parallel to lattice axes of the triangular lattice: (i) $V_{\text{wall}}(\boldsymbol{r}) = \varepsilon_{\text{wall}}(\sigma/|x - x_{\text{wall}}|)^{10}$ for a particle at position $\boldsymbol{r} = (x, y)$, where the x-direction is chosen perpendicular to the boundaries, while y runs parallel to the boundaries. The positions x_{wall} of the two walls are a distance D apart, $D = na\sqrt{3/2}$, n being an integer, and a being the lattice spacing of an ideal triangular lattice compatible with the chosen density. Various values of $\varepsilon_{\text{walls}}$ were studied [6] but here only data for $\varepsilon_{\text{wall}} = 0.0005$ will be given. (ii) Structured walls causing a periodic corrugation of the potential were created by choosing two rows of particles (running in the y-direction) fixed in the positions of this ideal triangular lattice. These fixed particles interact with the mobile particles with the same potential $V(r)$ as specified above. Summing up these potentials due to the fixed particles defines the corrugation potential V_{struc} of such structured boundaries.

21.2 Results of Monte Carlo Simulations

As it is standard, Monte Carlo (MC) moves are chosen by random selection of single particles, attempting a small random displacement in a square of linear dimension $\alpha = 0.206$ centered at the old position of the particle. Typical runs were performed for systems containing between $N = 20 \times 20$ and 60×60 particles, but also elongated geometries (e.g. 20×100) were studied, carrying out 10^6 MC steps per particle (MCS) in each run. In the y-direction parallel to the walls, periodic boundary conditions (pbc) were used. For the sake of comparison, also runs for "bulk" systems (with no walls and pbc in both directions) were done. Figure 21.1 shows configurations of the particles adjacent to the left wall. While the structured wall enhances the order near the wall, the particles are better localized near their ideal positions, the planar wall created a long range anisotropy in the mean square displacement of the particles! The particles move preferentially along the $d = 1$ rows parallel to

Fig. 21.1. Configurations of the particles in the first 9 rows adjacent to the left wall, for the structured wall (*left*) and the planar wall (*right*). 1000 configurations of a run lasting 10^6 MCS are superimposed, fixing the center of mass of the mobile particles in the same position

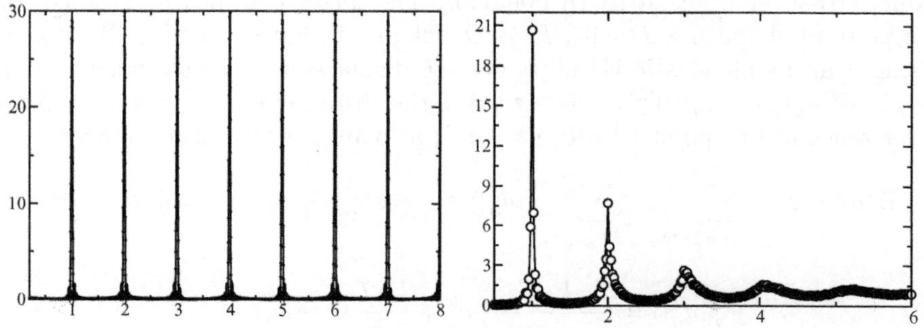

Fig. 21.2. Static structure factor $S(q)$ plotted vs. $qd/2\pi$ for structured walls (*left*) and planar walls (*right*). All data are for systems of 900 particles. For the planar walls the full curve is a fit to (21.1), adjusting the parameter A

the planar walls. While near both walls orientational order is enhanced (this remark can be made quantitatively by analyzing the appropriate order parameter [6]), positional long rang order (LRO) is enhanced by the structured wall but destabilized by the planar wall. This is seen by an analysis of the structure factor $S(q)$, where the wavevector $\boldsymbol{q}$ is oriented in the y-direction, $S(q) = \left\langle \sum_{\ell,\ell'} \exp[iq(y_\ell - y_{\ell'})] \right\rangle / N$, sums over ℓ and ℓ' being extended over all particles. Figure 21.2 shows that for the structured wall sharp Bragg peaks result while for the structureless repulsive boundary a typical fluid-like structure factor results (but with much higher first two peaks!) In fact, this structure factor is almost in quantitative agreement with a fit to the $S(q)$ for $1d$ harmonic chains [5]

$$S(q) = \sinh(A^2q^2/2)/[\cosh(A^2q^2/2) - \cos(qd)] \quad , \tag{21.1}$$

d being the distance between lattice axes in y-direction and A is a (fit) parameter which controls the width of the peaks (in $1d$ the width continuously goes to zero as $T \to 0$ and hence a crystal at $T = 0$ is restored [5]). Note that in Fig. 21.2 we deal with a system of 30 rows confined between two boundaries, rather than a true $d = 1$ system, and the state of the system is *not like a fluid* (then one would see lots of dislocations in the configuration picture, cf. Fig. 5 of [8]) but rather a "$2d$ smectic", a phase with orientational LRO but no positional LRO. Thus it would be wrong to consider the effect of planar walls as surface-induced melting. A planar wall enhances the mean-square displacement of two particles which are n lattice parameters apart in y-direction, as compared to the bulk, similar as at a free surface of an XY-model a faster power law decay of the spin-spin correclation occurs [9]; but in the x-direction normal to the wall the ordering tendency is enhanced. This is seen quantitatively from the density distribution $\rho(x)$, Fig. 21.3a. However, this effect of walls to make fluctuations anisotropic is only one ingredient causing the destruction of order in strips of width D: The second ingredient is the dimensional crossover from $2d$ to $1d$ behavior. The latter can also be studied for a system of shape $L \times D$ with $L \gg D$ but pbc in both x- and y–directions. Using a harmonic elastic Hamiltonian the displacement correlation function $B(y) \equiv \langle [u_y(y) - u_y(0)]^2 \rangle$, where $\boldsymbol{u}(y)$ is the displacement vector away from a reference lattice point $(x = 0,\ y = ma)$, m being an integer, becomes

$$B(y) = 2 \sum_{q_x = 2\pi n_x/D} \sum_{q_y = 2\pi n_y/L} \langle |u^2_{\boldsymbol{q}}| \rangle [1 - \cos(q_y y)] \quad , \quad \hat{q}_y = q_y/q. \tag{21.2}$$

$$\langle |u^2_{\boldsymbol{q}}| \rangle = \frac{k_B T}{(\lambda + 2\mu) q^2} \hat{q}^2_y + \frac{k_B T}{\mu q^2} (1 - \hat{q}^2_y) \quad . \tag{21.3}$$

Here n_x and n_y are integers and the sums cover all points within the first Brillouin zone, but omitting its origin (which would yield a uniform displacement of the whole crystal). The Lamé coefficients λ and μ are those of a bulk solid at the same density corrected for the overall hydrostatic pressure [8, 10]. From (21.2), (21.3) one can show that [6] $B(y) \propto \ln y$ for $a \ll y \ll D$ while [6] $B(y) \propto y/D$ for $D \leq y \ll L$ (Fig. 21.3b, inset).

21.3 Discussion and Concluding Remarks

In this work it was demonstrated that confinement of $2d$ crystals by external boundaries into quasi $1d$ strips of width D destroys the $2d$ positional quasi LRO {while the behavior $B(y) \propto \ln y$ destroys the delta function singularities of $S(q)$ at the points of the reciprocal lattice, power law singularities of $S(q)$ remain: this is meant by quasi–LRO [3,4]}. There is subtle interplay of finite size effects {crossover from $B(y) \propto \ln y$ to $B(y) \propto y/D$ at $y_c \approx D \ln(D/a)$ [6] see Fig. 21.3b)} and surface effects (smooth repulsive walls cause a layering effect, Fig. 21.3a, and a pronounced anisotropy of position fluctuations, Fig. 21.1). The confined crystal also exhibits very anomalous elastic properties: the bulk symmetry $C_{12} = C_{33}$ between elastic constants is broken [6], and one finds

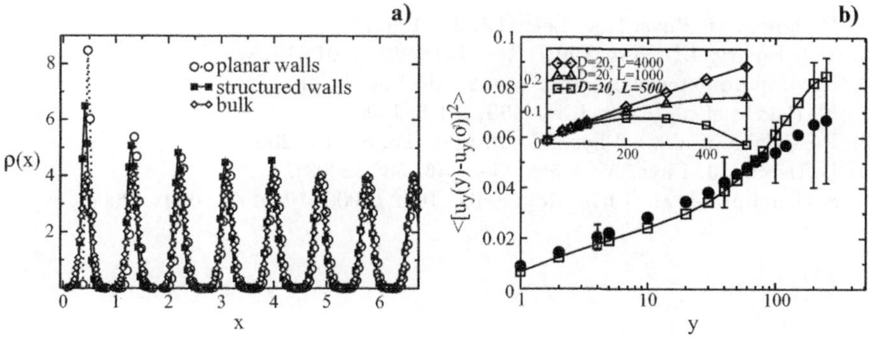

Fig. 21.3. (**a**) Density per row plotted vs. x, for systems of 3600 particles, analyzing 10^4 configurations taken from a run lasting 10^6 MCS. Only the first 6 rows adjacent to a flat planar wall at $x_{\text{wall}} = 0$ are shown, comparing results for both types of walls to the bulk behavior (for the structured wall, the first row of fixed particles is at $x_{\text{walls}} = -(a/4)\sqrt{3}$). (**b**) Comparison of $\langle [u_y(y) - u_y(0)]^2 \rangle$ vs. y (in logarithmic scale), as obtained from the simulations of a 20×500 system with pbc (*filled circles*) with the results of the corresponding harmonic theory (*squares*). The Lamé coefficients ($\lambda = 42$, $\mu = 41$, in units of $k_B T/\sigma^2$) are those of the bulk solid at the same ρ and T, obtained as described in [8,10]. *Inset* shows a linear-linear plot of $B(y)$ vs. y according to the harmonic theory for three choices of L (note that the pbc imply $B(y) = B(L - y)$ and hence $B(y) \propto y$ can be observed for $y \ll L$)

a vanishing shear modulus [6]. As a consequence, a 2d "crystal" confined by smooth repulsive walls is not really a crystal, but a kind of 2d "smectic liquid crystal" phase! In contrast, suitably corrugated walls enhance positional LRO, leading to sharp Bragg peaks (Fig. 21.2, left part). These results illustrate the sensitivity of (quasi) LRO to the boundary conditions of the system.

Finally, we believe that these phenomena should be observable in experiments with colloidal systems under suitable confinement.

Acknowledgement

We are grateful to the Deutsche Forschungsgemeinschaft for support (SFB TR6/C4), thank the NIC for computer time.

References

1. K. Zahn et al., Phys. Rev. **82**, 2781 (1999); *ibid.* **90**, 155506 (2003); K. Zahn, G. Maret, Phys. Rev. Lett. **85**, 365 (2000).
2. Q.H. Wei et al., Science **287**, 625 (2000); C. Lutz et al., J. Phys.: Condens. Matter **16**, S4075 (2004).
3. D.R. Nelson, in: *Phase Transition and Critical Phenomena*. Vol. 7. C. Domb, J.L. Lebowitz (Eds.) (Academic, London 1983).

4. B. Jancovici, Phys. Rev. Lett. **19**, 20 (1967).
5. V.J. Emery, J.D. Axe, Phys. Rev. Lett. **40**, 1507 (1978).
6. For a more extensive report, see A. Ricci et al., preprint
7. K. Bagchi et al., Phys. Rev. E**53**, 9794 (1996).
8. P. Nielaba et al., J. Phys.: Cond. Mat. **16**, S4115 (2004).
9. B. Berche, J. Phys. A.: Math. Gen. **36**, 585 (2003).
10. S. Sengupta et al., Phys. Rev. E **61**, 1072 (2000); *ibid.* **61**, 6294 (2000).

List of Contributors

SPRINGER PROCEEDINGS IN PHYSICS

90 **Computer Simulation Studies in Condensed-Matter Physics XV**
Editors: D.P. Landau, S.P. Lewis, and H.-B. Schüttler

91 **The Dense Interstellar Medium in Galaxies**
Editors: S. Pfalzner, C. Kramer, C. Straubmeier, and A. Heithausen

92 **Beyond the Standard Model 2003**
Editor: H.V. Klapdor-Kleingrothaus

93 **ISSMGE**
Experimental Studies
Editor: T. Schanz

94 **ISSMGE**
Numerical and Theoretical Approaches
Editor: T. Schanz

95 **Computer Simulation Studies in Condensed-Matter Physics XVI**
Editors: D.P. Landau, S.P. Lewis, and H.-B. Schüttler

96 **Electromagnetics in a Complex World**
Editors: I.M. Pinto, V. Galdi, and L.B. Felsen

97 **Fields, Networks, Computational Methods and Systems in Modern Electrodynamics**
A Tribute to Leopold B. Felsen
Editors: P. Russer and M. Mongiardo

98 **Particle Physics and the Universe**
Proceedings of the 9th Adriatic Meeting, Sept. 2003, Dubrovnik
Editors: J. Trampetić and J. Wess

99 **Cosmic Explosions**
On the 10th Anniversary of SN1993J (IAU Colloquium 192)
Editors: J. M. Marcaide and K. W. Weiler

100 **Lasers in the Conservation of Artworks**
LACONA V Proceedings, Osnabrück, Germany, Sept. 15–18, 2003
Editors: K. Dickmann, C. Fotakis, and J.F. Asmus

101 **Progress in Turbulence**
Editors: J. Peinke, A. Kittel, S. Barth, and M. Oberlack

102 **Adaptive Optics for Industry and Medicine**
Proceedings of the 4th International Workshop
Editor: U. Wittrock

103 **Computer Simulation Studies in Condensed-Matter Physics XVII**
Editors: D.P. Landau, S.P. Lewis, and H.-B. Schüttler

104 **Complex Computing-Networks**
Brain-like and Wave-oriented Electrodynamic Algorithms
Editors: I.C. Göknar and L. Sevgi

105 **Computer Simulation Studies in Condensed-Matter Physics XVIII**
Editors: D.P. Landau, S.P. Lewis, and H.-B. Schüttler

106 **Modern Trends in Geomechanics**
Editors: W. Wu and H.S. Yu

107 **Microscopy of Semiconducting Materials**
Proceedings of the 14th Conference, April 11–14, 2005, Oxford, UK
Editors: A.G. Cullis and J.L. Hutchison

108 **Hadron Collider Physics 2005**
Proceedings of the 1st Hadron Collider Physics Symposium, Les Diablerets, Switzerland, July 4–9, 2005
Editors: M. Campanelli, A. Clark, and X. Wu

109 **Progress in Turbulence II**
Proceedings of the iTi Conference in Turbulence 2005
Editors: M. Oberlack, G. Khujadze, S. Guenther, T. Weller, M. Frewer, J. Peinke, S. Barth

110 **Nonequilibrium Carrier Dynamics in Semiconductors**
Proceedings of the 14th International Conference, July 25–29, 2005, Chicago, USA
Editors: M. Saraniti, U. Ravaioli

111 **Vibration Problems ICOVP 2005**
Editors: E. Inan, A. Kiris

112 **Experimental Unsaturated Soil Mechanics**
Editor: T. Schanz

GPSR Compliance

The European Union's (EU) General Product Safety Regulation (GPSR) is a set of rules that requires consumer products to be safe and our obligations to ensure this.

If you have any concerns about our products, you can contact us on ProductSafety@springernature.com

In case Publisher is established outside the EU, the EU authorized representative is:

Springer Nature Customer Service Center GmbH
Europaplatz 3
69115 Heidelberg, Germany

Batch number: 10755053

Printed by Printforce, the Netherlands